Red Shirts and Leather Helmets

Merry Christmas 1984
to
Toby & Sue Hardy
Sincerely
Steve Frady

RED SHIRTS and LEATHER HELMETS

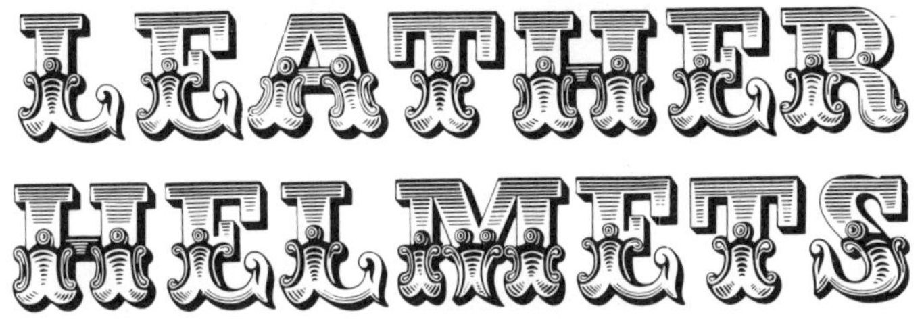

Volunteer Fire Fighting on the Comstock Lode

Steven R. Frady

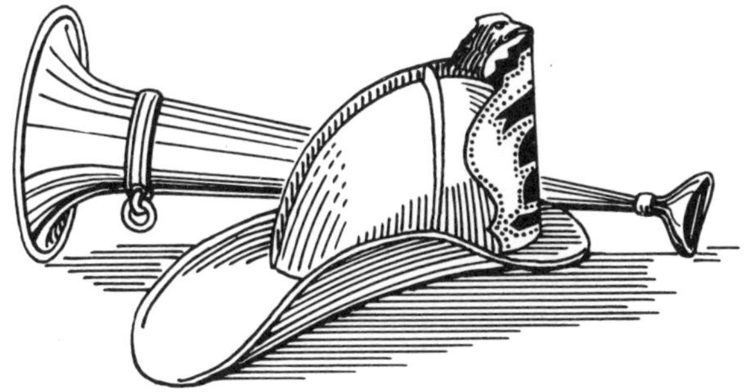

University of Nevada Press
Reno 1984

Library of Congress Cataloging in Publication Data

Frady, Steven R., 1948–
Red shirts and leather helmets.

Includes bibliographical references and index.
1. Fire-departments – Nevada – Virginia City – History.
2. Comstock Lode (Nev.) – Fires and fire prevention – History, I. Title.
TH9505.V57074F73 1983 363.3′78′0979356 84–2335
ISBN 0-87417-086-9 (cloth)
ISBN 0-87417-087-7 (paper)

UNIVERSITY OF NEVADA PRESS, RENO, NEVADA 89557 USA

BOOK DESIGN BY DAVE COMSTOCK
PRINTED IN THE UNITED STATES OF AMERICA

Contents

	Introduction	vii
I.	From Snowballs to Hand Pumpers	1
II.	Dangerous Work	24
III.	Gold Hill's Great Fire of 1870	33
IV.	The Social Life of a Fireman	39
V.	Home Sweet Home	57
VI.	Scandal	69
VII.	Rivalries and Friendships	76
VIII.	Man the Brakes!	92
IX.	The Hooks	108
X.	Jump Her Lively, Boys!	115
XI.	Fighting Fire With Fire	129
XII.	A Deed Most Foul	140
XIII.	A Love-Hate Affair	146
XIV.	Fire in the Mines	154
XV.	Conflagrations	162
XVI.	Virginia City's Great Fire of 1875	170
XVII.	Bitterness and a Paid Department	184
	Appendix 1. Chronology of the Comstock Volunteer Fire Companies	222

Appendix 2. Constitution and Bylaws of Eagle Engine Company No. 3, Virginia, Nevada, 1865 224

Appendix 3. Virginia City's Great Fire, October 26, 1875 236

Appendix 4. Glossary 238

Notes 243

Index 255

Sidebars

Alf Doten: Editor in a Leather Helmet 20

The Death of "Old Butt" 31

Chief John Marks: Acts of Sedition 73

Chief Tom Peasley: A Tough Man 87

Fireman Dave Tweedie: Making the Daily Run 219

Introduction

For millions of children growing up in America, there was no greater thrill than watching the volunteer firemen of their town scurry up the street with their fire engines at the sound of the alarm bell. There was something exciting about those early firemen, especially when they wore their red shirts and leather helmets on formal occasions. Strong men and little boys alike longed to join their ranks as they marched by in the Fourth of July parade.

No matter what the emergency, the firemen were always on hand to help. In the shock, smoke, and confusion of a conflagration, when forced out of a warm bed in the dead of night by the town's alarm bell or whistle, the example set by these volunteers was one of leadership, heroism, and sacrifice. In times of fire or other emergency, the townspeople looked to their firemen for cool-headedness and courage, which saved lives that otherwise would have been lost.

Americans have had ample opportunity to see their heroes in red shirts and leather helmets at work. The country was plagued from the beginning by the ravages of fire. Wholesale destruction was visited indiscriminately upon large cities, small communities, villages, and even transient tent towns. Candles, oil lamps, and lanterns, side-by-side construction of endless blocks of wooden buildings, arsonists, and accidents combined to create calamities that often struck more than once. Major fires occurred in Boston, San Francisco, New York, Chicago, and more than a dozen of the nation's larger cities, not to mention hundreds of smaller cities and towns. It has been said that the 1870s were the most destructive years in terms of fire loss in American history.

From 1860 to 1880, Nevada's cities, towns, and fledgling mining camps were repeated victims of the "fire fiend," as early newspapermen labeled blazes. In 1865, several of Washoe City's largest buildings were destroyed by fire, and an arsonist was successful in destroying an entire business block in Carson City, with a reported loss of more than $25,000. In 1872, Pioche was swept by a fire that destroyed twelve buildings at a loss of $50,000. One year later, the tiny town of Genoa lost six of its principal buildings in a blaze. The conflagration that destroyed the business district of Treasure Hill, in 1874, was so devastating that the business section of that town was never rebuilt.

A second, even more destructive fire swept through Pioche in 1876; and in 1877, the main business section of Humboldt Wells was levelled by the work of an arsonist. The year 1879 brought destruction by fire to three major Nevada towns. In Eureka, two thousand were left homeless and half the town reduced to ashes. In Reno, hundreds were left homeless, six were killed, and more than $900,000 in property was lost. Tuscarora's Grand Prize Mill, Wilkins' Hotel, and a number of other buildings were destroyed in a fire that same year. The destruction of several buildings in the small town of Sutro, near Virginia City, in 1879, prompted that town to begin serious efforts to establish a fire department.

Throughout the twenty-year period from 1860 to 1880, Gold Hill and Virginia City experienced dozens of fires. Again and again, fire wiped out whole residential areas, and laid waste to entire business blocks. Each time the wave of destruction swept through the Comstock, the area was rebuilt to prosper once more. Each time the fire alarm sounded, volunteer firemen scrambled from the various sections of town to the engine houses, to get the hand pumpers and hose carriages to combat the blaze.

From 1861 to 1866, six engine companies, with accompanying hose companies, and a hook and ladder company were formed in Virginia City. They became the nucleus of the Virginia Fire Department. In Gold Hill, the first hose company was formed in 1863, followed by two other hose companies in 1866. Two of the three original hose companies were later reorganized as engine companies, one with a hand pumper, the other with a new steam fire engine. With both engine companies came the formation of sister hose companies, thus further swelling the ranks of the fire department of Gold Hill.

The volunteer firemen of Virginia City and Gold Hill were members of a system whose origins and operations can be traced to the first American fire company, organized by Benjamin Franklin in Philadelphia. The later companies followed Franklin's organizational philosophy, which called for only thirty to forty members in a company, and thus led to the formation of many fire companies and, quite often, bitter rivalries. The early fire companies were not unlike their descendants, with prominent citizens and civic leaders in the lead, and the working class filling out the company roll. In Franklin's era, such notable citizens as George Washington, John Jay, Thomas Jefferson, Samuel Adams, Alexander Hamilton, Paul Revere, John Hancock, Aaron Burr, and even Benedict Arnold, were members of volunteer fire companies. These early groups introduced many of the fine traditions of the fire service of America.

One of these traditions was patriotism. With men like Washington, Franklin, and Jefferson among the first fire leaders, it is no wonder that patriotism remains one of the foundation blocks of the modern American fire service. The organizers of the first fire companies on the Comstock were also patriots. Men like Thomas Peasley and John Van Buren Perry—quick to defend the Union with fists or pistols, if necessary—led miners, hotel keepers, teamsters, laborers, and saloon keepers in battling the "fire fiend" on the Comstock Lode, while war was waged farther east to determine the future of the nation.

It was to men such as Peter Larkin, John Marks, and James K. B. "Kettlebelly" Brown that Comstock residents looked for security from the ravages of fire. It was these men and many more who braved inclement weather, flame, smoke, and crumbling walls to protect the mining towns from fearful disasters like those suffered in New York and San Francisco on so many occasions. Although they did the best they could with the technology of the day and the water resources at hand, fire and politics put an end to their way of fighting fire. The old Virginia Fire Department was eliminated in 1877, paving the way for a paid fire department. Gold Hill resisted cuts in budgets, efforts to reorganize, and other pressures, to retain its volunteer fire-fighting force until Liberty Engine Company finally succumbed to modernization and disbanded in 1938.

The story of the fire companies and the firemen of the early Comstock Lode has largely been ignored. Discussion has been limited to sentences, or in some cases to a generous paragraph in the history books. Their contribution to the traditions of the Comstock and of Nevada, their pioneering role in the fire service of the Silver State, their individual battles with fire (and sometimes with each other), and their many acts of heroism have largely been forgotten, although other aspects of the development of Gold Hill and Virginia City have been addressed at length.

This story of the volunteer firemen of Virginia City and Gold Hill has taken more than five years to document. It has been an enormous task, made difficult by the relative scarcity of records. Those in existence are scattered throughout several states in public and private collections, but the search for manuscripts and photographs has been a rewarding one. Uncovered during this quest have been such important records as the Grand Register of the Virginia Fire Department, listing all members of the organization from 1861 to 1875. The book was forgotten in a vault, but has now been turned over to the Comstock Firemen's Museum in Virginia City. Also uncovered, in yet another vault, were the records of the Gold Hill Fire Department from 1867 to 1889, minutes of Liberty Engine Company No. 1 meetings, and the Roll Book of the Gold Hill Fire Department, all of which were presented to the Comstock Firemen's Museum by former Storey County Assessor Jack Flannagan. Also found were original handwritten copies of the rules of the Virginia Paid Fire Department, as well as resolutions establishing fire districts and locations for hose houses and the Corporation Fire House after Virginia City's Great Fire of 1875.

A large number of photographs, many of them unpublished, have also been uncovered. These include a view of the Sutro Fire Department proudly displaying their hand pumper, formerly the Knickerbocker No. 5 of the Virginia Fire Department, and a rare action shot of a Virginia City fire. Portraits of many of the firemen of Virginia City and Gold Hill have also been discovered, including one of Virginia City's first fire chief, Thomas Peasley. There are also photographs of the engine companies and their equipment, and many others that help depict the life of a volunteer fireman on the early Comstock.

Many sources provided the material for this book, and they are listed in the bibliography. Three sources deserve mention here because of their assistance in putting together the story of the Comstock's fire service. *The Journals of Alfred Doten, 1849–1903,* contained critical dates and incidents, and also gave a colorful picture of Virginia City and Gold Hill. Alf Doten was a newspaper reporter and editor, and a member of Washoe Engine Company No. 4. This gave him an excellent vantage point from which to observe the growth of the fire departments of Gold Hill and Virginia City from volunteer to paid organizations, as well as the development of the Virginia Exempt Firemen's Association, of which he was also a long-time member. (For definitions of such early-day terms as "exempt," see Appendix 4, "Glossary.")

Because Doten was both editor of the *Gold Hill News* and a member of the fire department, his reporting might be suspect. However, his reports of fire department activities seem to be as complete, if not more so, than those of his colleague, Joseph Goodman, editor of the *Territorial Enterprise*, who maintained his membership in Virginia Engine Company No. 1 for only one year. By comparison to the *Gold Hill News*, the *Territorial Enterprise* shows little concern for the day-to-day operations of the fire departments of Virginia City and Gold Hill, often reporting about them only upon the occasion of a major fire or internal incident. Making the task of sifting information about the fire companies more difficult, and the corroboration of stories sometimes impossible, is the lack of a complete set of the *Territorial Enterprise*. Ironically, these papers were burned in the Great Fire of 1875, along with most of the records of the various companies which made up the Virginia Fire Department.

Adding to the research problems were the many legends surrounding the fire department, its members, and the Great Fire of 1875. Contradictions and inaccuracies in sources—such as Thompson & West's *History of Nevada, 1881,* which indicated in one section that the fire department had been burned out in the Great Fire, while only paragraphs later stating that the same equipment supposedly destroyed was presented to the Exempts—were another problem that required hours of research.

Putting the story together was like building a giant jigsaw puzzle, with some pieces appearing and seeming to fit, but always missing that critical corner—that critical piece of information. Sometimes leads tapered off into dead ends. Sometimes, however, the reward of perseverance paid off, with answers materializing almost miraculously. With all of the research, however, questions still remain, such as the fate of the Comstock fire equipment in the years following the Great Fire of 1875. There is also the question of where the artifacts collected by the Exempts eventually wound up. While some have surfaced in the collections of the Nevada Historical Society and Warren Engine Company No. 1, the majority of the items have seemingly vanished.

* * *

The difficult task of sifting through countless records, newspapers, photographs, and other documents was made easier by a number of people and organizations. For their assistance, I offer sincere thanks. Among the most helpful was historian and author Richard C. Datin, whose encouragement, research assistance, and good-natured kidding were invaluable. Others deserving of recognition include: Warren Engine Company No. 1 of Carson City, whose archives were opened for me, and whose members have been a constant source of encouragement; the staff of the Nevada Historical Society, who answered without complaint my endless requests for this record, that box of materials, or yet another collection of photographs; Tim Gorelangton and the staff of Special Collections, Library, University of Nevada, Reno, without whose assistance many photographs would not have been located and who provided countless research ideas and leads; Chief Bernard Sease of the Carson City Fire Department, who provided technical assistance with the photograph collections and whose love of the old volunteer fire companies helped encourage me to continue with the project; Storey County Clerk Marlene Andreasen, who put up with probing questions about the records in her vault, and who helped me search through stacks of books and papers for more information, documents, and leads; the Nevada State Library, whose staff assisted with weeks of research in the microfilmed records of the Comstock newspapers; Ormsby County Librarian Virginia Rule, for her special assistance at critical times; the Storey County Volunteer Fire Department – Virginia City District, a new Liberty Engine Company No. 1, and the Comstock Firemen's Museum for invaluable photographs and records; Bob Laxalt, Nick Cady, and Rick Stetter of the University of Nevada Press, for their encouragement, assistance, and expertise in the preparation of the manuscript; countless firemen and other friends who have offered words of encouragement for the project; and my wife, Inez, who has put up with papers, books, photographs, and artifacts spread around the house, as well as having to listen to hours of historical trivia dug up during research. Her patience during the research and writing of this book has been of critical importance to its successful completion.

I. From Snowballs to Hand Pumpers

They wore red shirts and leather helmets, and their battle cry was "First water!" They were the volunteer firemen of the Comstock and they occupied a special position in the society of the mountainside mining towns of Virginia City and Gold Hill.

With the discovery of gold and later silver on the slopes of Mt. Davidson, the rush to Washoe was on. Miners, promoters, speculators, saloon keepers, merchants, ex-soldiers, and others began to stream into Virginia City and Gold Hill. The rush caused a mass exodus from the gold fields of California, and uprooted thousands of pilgrims from the East, who left their homes hoping to strike it rich in Virginia City.

In the fall of 1859, Virginia City's first street was laid out, and along it were hastily erected shanties, tents, and other structures. Mark Twain wrote:

> Virginia had grown to be the "livest" town, for its age and population, that America had ever produced. The sidewalks swarmed with people—to such an extent, indeed, that it was generally no easy matter to stem the human tide. The streets themselves were just as crowded with quartz wagons, freight teams and other vehicles. The procession was endless. So great was the pack that buggies frequently had to wait half an hour for an opportunity to cross the principal street.[1]

Virginia City was growing up and growing wild. In *Roughing It*, Twain painted a picture of a town with brass bands, banks, hotels, military companies, "hurdy-gurdy" houses, murders, inquests, riots, wide-open gambling palaces, "a whiskey mill every fifteen steps," a dozen breweries, half a dozen jails and station houses "in full operation, and some talk of building a church." Large, permanent structures of brick and stone were being erected and the wooden suburbs were spreading out in all directions.[2]

Underground, another city was busily at work in the miles of tunnels, drifts, adits, galleries, and cooling rooms. The mines were flourishing. Fortunes were being made by the men who controlled the mills, the mines, and, later, the railroads and commercial enterprises supplying the Comstock with the necessities of life.

Wages for the miner to spend ranged from four to six dollars a day, and spend he did. There was hardly a hurdy-gurdy, saloon, or brothel that wasn't packed nightly, as the miners spent what was left of their hard-earned wages after necessities were purchased. Even the necessities were expensive. The first barrel of flour brought into the camp was sold at auction, with a minimum bid of one dollar per pound. Nails sold for the same amount, with shovels priced at nine dollars each. When John Moor opened the first mercantile store, he sold blankets at two dollars a pair, brandy at six dollars a gallon. A sleeping place in his back room, with a blanket, cost a minimum of one dollar per night. Such accommodations usually meant a small space on a hard, drafty floor, shared by so many that the snoring of the sleeping men sounded like a swarm of irate bees. In addition, there was always that little extra that came with such accommodations at no charge—the bedbug.[3]

The Virginia City of 1860 was described by one of its citizens:

> Frame shanties pitched together as if by accident; tents of canvas, of blankets, of brush, of potato sacks and old shirts, with empty whisky barrels for chimneys; smoking hovels of mud and stone; coyote holes in the hillsides forcibly seized by men; pits and shanties with smoke issuing from every crevice; piles of goods and rubbish on craggy points, in the hollows, on the rocks, in the mud, on the snow, everywhere scattered broadcast in pellmell confusion.[4]

The city had a slant to it like the pitch of a roof, Mark Twain wrote. Each street was a terrace, and from each to the next street below was a descent of forty or fifty feet. Fronts of houses were level with the streets they faced, but the rear first floors were often propped on stilts, or built on top of the building facing the next street down. "It was a laborious climb, in that thin atmosphere," Twain wrote, "to ascend from D to A Street, and you were panting and out of breath when you got there; but you could turn around and go down again like a house afire — so to speak."[5]

In this atmosphere there was a twenty-four-hour turmoil that mixed the sounds of the hoisting works bringing ore to the surface, bells ringing as the cages moved miners to their work below the city, and great steam-driven machinery working creaking cables over pulley wheels, interrupted by the sound of a check valve popping off, or the dull thud of a shot of Giant Powder somewhere deep in the bowels of the city. Streets were jammed with giant wagons pulled by teams of oxen or horses, on their way to the mills along the Carson River, or bringing freight into town. Buggies, horses, and pedestrians made their way through clogged sidewalks and muddy streets filled with oxen sleeping in the midday heat, chickens scrambling, and Chinese talking and shouting in a chant-like language that everyone else somehow found amusing.

Virginia City was bustling with signs of prosperity and permanence. Under construction were schools, churches, substantial commercial hardware and grocery stores, furniture and clothing outlets, and fine butcher shops. Boardwalks soon replaced the muddy walkways along the streets, and throughout the town was the confident expectation of a rich and prosperous future for the largest town in the territory. It was a wonderful, confused, and crazy place in which to live, filled with strange sights and sounds.

And it was a firetrap.

One of the first fires to occur in Virginia City was at a cabin on A Street in 1861. Although there was no organized fire department, the nucleus of the future Virginia Fire Department was to be found among the citizens gathered before the burning cabin, pelting it with snowballs that cold day in February. The fire fight deteriorated into a snowball melee, but as a result of that blaze, efforts were soon underway to form first a bucket brigade, and then Nevada's first fire engine company, Virginia Engine Company No. 1.[6]

The necessity for an organized fire-fighting force on the Comstock was obvious. Most structures in the town were constructed of wood, with cloth or paper covering interior walls. Buildings shared common walls and often chimneys, and long rows of wooden structures could act as a bridge for the sweep of flames during a conflagration. Chimneys were made of wood, mud, barrels chinked with mud, or other materials, often resulting in fires on neighboring roofs, on tent walls, in rubbish piles, and in the surrounding sagebrush. Illumination at night was provided by

Virginia City in 1861 was a firetrap. There were several substantial brick buildings, but even they proved not to be "fireproof" during major conflagrations. This view looks north along C Street. (Special Collections, University of Nevada, Reno, Library)

a miner's candlestick poked into wooden walls. Candles often burned through and dropped onto the floor, catching bedding, curtains, walls, and other materials on fire. Coal oil lamps burst from extreme heat and rough use, spreading fire rapidly along tinder-dry floors.

The constant threat of a major fire, emphasized by an occasional small blaze, was the catalyst for the formation of Virginia Engine Company No. 1 and Nevada Hook and Ladder Company No. 1 in 1861. Through 1866, five more engine companies organized to form the Virginia Fire Department, and in the 1870s, hose companies were organized to run with the town's six engine companies. In Gold Hill, efforts to form a fire company began in 1863, and by 1868, the Gold Hill Fire Department boasted an engine company and three hose companies. (See Appendix 1.)

The fire companies of the Comstock were comprised of laborers from the depths of the mines, teamsters, saloon keepers, store owners, newspaper reporters, railroad officials and bank officers. But when they wore the special red shirt of volunteer firemen, they were equals in society. Side by side, they faced the same dangers in trying to defeat the "fire fiend."

The broad background and varied social and economic status of the members of the companies of the Virginia Fire Department is reflected in the membership roll of Virginia Engine Company No. 1 in 1862. A sampling of the members reveals a broad slice of the Comstock society: I. C. Bateman, owner of the International Hotel; G. A. Fogg, proprietor of the Washington Market; Thomas Andrews, undersheriff; George I. Lammon, stock broker; William H. Barstow, store clerk; James Stapleton, laborer at the Zouave Mining Company; J. M. Walker, miner at the Ophir; Terence McGinnis, brick mason; George Ennis, druggist; Martin Groestta, Virginia Saloon; R. T. Smith, miner; John Reed, hatter; John Smith, laborer at the Mexican Mill; A. Hirschman, jeweler; Charles Fish, district recorder; Sam Paster, proprietor of Paster's Saloon; William Milligan, miner; George May, deputy sheriff; Sam Wetherill, acting judge; I. E. Brokaw, Second Ward policeman; George Shaw, Second Ward alderman; James Waters, miner at the Ophir; R. C. Hardy, proprietor of the Niagara Saloon; M. C. Hillyer, miner.[7]

One had to be voted a member of a fire company, insuring a certain exclusivity in the ranks. A proposed member's name was submitted to an investigating committee, which reported back to the general membership upon the qualifications of the prospect. If the report were favorable, the prospect was voted on by the company, in secret ballot. Five black balls were enough to disqualify a man for membership. Once elected, the new

Virtually every segment of society from bankers to laborers was represented in the fire companies on the Comstock Lode. These men were members of Yellow Jacket Hose Company No. 2 of the Gold Hill Fire Department in the early 1860s. (Comstock Firemen's Museum)

member was required to sign the constitution of the company and pay a two-dollar fee, plus the cost of a membership certificate.[8]

Being a prominent citizen of the town virtually insured membership in one of the fire companies. Among the firemen were: state senators John Piper and James Phelan, U.S. Senator John Jones, newspaper editors Joseph Goodman, Alf Doten, and Wells Drury, aldermen A. V. Comstock and Jacob Young, Jr., mayors J. S. Kaneen and G. W. Hopkins, Judge and later Postmaster D. O. Adkison, mining engineer Phillip Deidesheimer, and many others of high social position on the Comstock. Several law enforcement officers also wore the red shirt of a fireman, including five chiefs of police, four town marshals, several undersheriffs, and numerous deputy sheriffs and city policemen.[9]

Fire department members were allowed to wear the colorful uniforms of their respective companies, which resembled fancy dress military uniforms. These were often trimmed in white or blue, with large numbers, fancy letters, or a combination of both, designating the company to which the wearer belonged. The uniform was completed by the addition of blue or black pants and the traditional fireman's hat or helmet, upon which was a giant shield bearing the name and number of the company. Around their waists, the volunteer firemen wore belts with giant buckles of Comstock silver or neatly worked leather, bearing the number of the company, which was sometimes surrounded by a fancy design of firemen's tools. Besides being part of the dress uniform for fancy dress balls, parades, funerals, and other formal occasions, the belt served the useful purpose of carrying tools and identifying the men and companies responding to fire alarms. Anyone found impersonating a fireman or wearing the badge of a fireman could find himself facing time in the city jail with the prospect of a heavy fine.

Fancy certificates of membership, or firemen's commissions, were issued to the members of the various companies. These were often emblazoned with the company's motto and a drawing of the company's engine or hose carriage. In addition, members also received, upon payment of a fee, a certificate of membership in the fire department.

The certificate of the Virginia Fire Department, created in 1865, was "a very beautiful one," according to the *Gold Hill News*. The design featured a robed female, representing Virginia City, holding a wand in her hand. Over her head was a sunburst, and directly beneath, at the bottom of the certificate, was a view of the city. On one side of the certificate was a quartz mill showing miners at work, with a border of scroll work incorporating various firemen's implements. Above the view of Virginia City was a spot for affixing the seal of the Virginia Fire Department. Various department officials, including the secretary, chief engineer, and president, signed the commission, which also bore the date of admission of the member. The certificate, which cost two dollars and fifty cents and offered proof of membership in claims of exemption from jury duty, "will constitute a beautiful ornament for a drawing room," the *Gold Hill News* stated.[10]

The certificate of membership in the Gold Hill Fire Department, unlike the black and white rendering of the Virginia Fire Department, was a full color lithograph, executed in 1878 and printed by H. S. Crocker of San Francisco. The certificate bore pictures of the hand engine of Liberty Engine Company No. 1, the steamer of Yellow Jacket Engine Company No. 2, and a hose carriage representing Lincoln Hose Company No. 3. Also shown were likenesses of the houses of the three fire companies, below which was a rendering of Upper Gold Hill, showing the many houses, businesses, and the famed Crown Point railroad trestle. In the upper half of the commission were various scenes portraying the dangers of a fireman's life as well as lifesaving heroics.[11]

Members of volunteer fire companies were governed by strict rules of conduct which called for expulsion, or at least a heavy fine, for such things as starting fights, failing to perform duties, or failing to maintain the proper decorum of a volunteer fireman. Also subject to expulsion from the company was someone who refused an officer's orders at a fire, who became delinquent in his dues, or who was found "repeatedly deficient in the discharge of his duty or does not exert himself to arrive at the engine or apparatus."[12]

Members and officers had specific duties at fires, which were spelled out in the constitution and bylaws (see Appendix 2) of each company. Each member was required to assist in pulling the apparatus to the fire, and then back to the engine house after the incident. The first arriving member was charged with acting as foreman of the company, taking charge of the foreman's trumpet, and directing the efforts of the company until relieved. Members were not allowed to leave the apparatus when on duty without the permission of the officer in charge. They were required to attend all meetings and drills of the company, as well as providing themselves with a company uniform within sixty days after becoming a member.

Officers were designated by white belts and helmets, and were often presented with special gold-mounted silver badges in the shape of a helmet shield as a symbol of their office. At fires, they had absolute

John T. Lee, second assistant engineer of Washoe Engine Company No. 4, in 1869. Note spanner wrench and nickel letters on belt. White helmet shield indicates his rank. (Special Collections, University of Nevada, Reno, Library)

Fireman's commission from the Gold Hill Fire Department. Only 500 of these certificates were printed; they were presented to members of the town's fire companies, who paid a fee to belong to the fire department, thus entitling them to certain privileges. (Author's collection.)

James Malone, a miner, in 1869. White helmet shield, white belt with title, and trumpet were symbols of office. (Warren Engine Company No. 1)

Officer's helmet shield from Neptune Hose Company No. 5. Such symbols of office were prized possessions, won in hard-fought elections. (Warren Engine Company No. 1)

power. The foreman of the company was charged with taking overall command at all drills, parades, and fires, and with insuring the good order and repair of the apparatus in his charge. The first assistant foreman was responsible for seeing that all equipment was well manned at fires, and for having the engine cleaned and repaired, if necessary, after fires. In the absence of the foreman, the first assistant was in command of the company. The second assistant foreman was in charge of the hose carriage and hose of the company, and saw to it that the carriage and hose were cleaned and properly stored after drill or fire.[13]

Elections for company or fire department offices usually generated considerable excitement for the volunteer firemen of the Comstock. When nominations were made, they received special attention in the local newspapers. And when extensive campaigns were conducted, they were reported with all the flair and detail of a presidential election. When department elections for chief and assistants drew near, the list of those firemen whose dues were current and who were not suspended, and were thus eligible to vote, were printed in the newspapers. Endorsements by the various fire companies of the candidates of their choice, many of whom were members of other companies, also received considerable attention in the newspapers. Sometimes the companies endorsed a candidate from a rival company, or a candidate nominated by another company, who was a member of yet a third company.

Elections were governed by the rules of the fire department. Polling places might be located in the engine houses, or commercial establishments such as Wood's Cigar Stand, where ballots were submitted for the election of 1867. Election judges were chosen by the Board of Fire Delegates, with representatives selected from several of the fire companies to act as a counting board and as referees at the polling places. Sometimes the services of referees were necessary, as Alf Doten wrote of the 1867 election: "Hugh Kerrin chosen Chief, J. C. Willock 1st Assistant, and S. D. (Wanie) Wright 2nd Engineers – Only 1 or 2 little fights – got along finely, and through counting & all right by 7 o'clock."[14]

Hard feelings between candidates and their supporters sometimes lasted for weeks after the elections. The 1865 election for chief of the Virginia Fire Department resulted in such ill feelings that the *Gold Hill News* remarked upon it: "Some of the defeated candidates may feel a little sensitive for awhile, but will, under the prompt and efficient guidance of the bully Chief and his able conferees, soon get the scalp healed of all election sores."[15] More often, however, elections were cause for grand celebrations. Even late into the century, in 1897, the *Territorial Enterprise* delighted in telling of yet another celebration sparked by the Gold Hill Fire Department election:

> After serenading the newly elected officers, the band, which was composed of fifteen pieces, went down to Liberty engine house, where the rigors of a stag dance were brightened by the presence of 75 gallons of beer and a sumptuous banquet consisting largely of crackers and cheese. While the dance was in progress a giant powder salute awoke the echoes of Gold Hill. The crowd which assembled in the hall became so great that the carts were run out into the street to accommodate the visitors. At least 250 men were present on the floor and everyone was kept moving by the dispensers of beer, who were going about with sprinklers overflowing with cheer. Hilarity was kept at a high ebb, some of the fire boys working off a little of their superfluous energy by hoisting a broom on the flag-staff of the engine house.[16]

The qualifications for chief of the fire department seem to have been minimal: a degree of popularity mixed with a seeming disregard for personal safety and a willingness to lead. From 1861 to 1876, twelve men served as chief of the Virginia Fire Department: Thomas Peasley, 1861–1862; Peter Larkin, 1863–1864; M. R. "Riff" Williams, 1865–

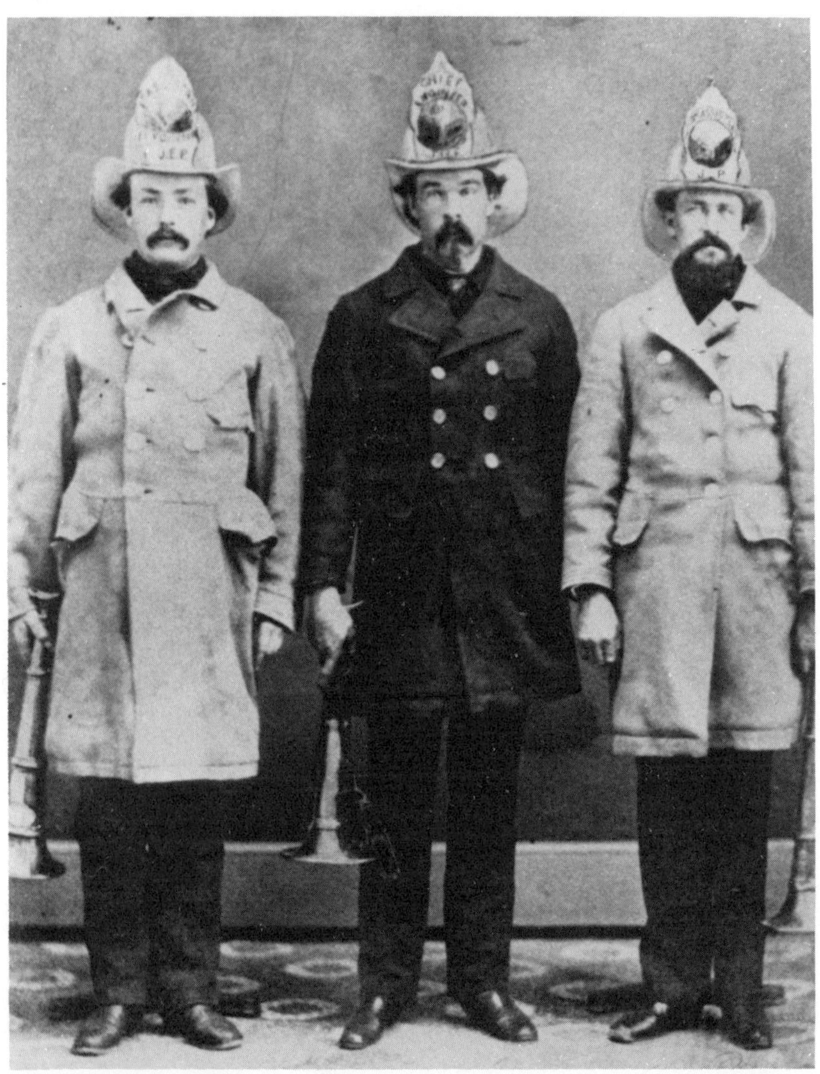

Officers of the Virginia Fire Department in 1869: James Parker of Monumental Engine Company No. 6, 1st Assistant; Thomas Fox, Washoe Engine Company No. 4, Chief; James Phelan, Young America Engine Company No. 2, 2nd Assistant. (Nevada Historical Society)

William Pennison of Knickerbocker Engine Company No. 5, in 1870, while serving as chief of the Virginia Fire Department. The heavy wool coat was "protective clothing" during a fire. Pennison later served seventeen years as chief of the Virginia Paid Fire Department after the volunteer companies disbanded. (Nevada Historical Society)

1866; Hugh Kerrin, 1867; J. C. Willock, 1868 (died in office, December 13, 1868; S. D. "Wanie" Wright was elected to serve as chief through the unexpired term); Thomas H. Fox, 1869; William Pennison, 1870; James Malone, 1871–1872; Joseph P. Bell, 1873; James K. B. "Kettlebelly" Brown, 1874; Frank McNair, 1875–1876.[17]

By 1881, only Kerrin, Fox, Pennison, Brown, and McNair were still alive. Pennison, a member of Knickerbocker Engine Company No. 5, and "Kettlebelly" Brown, a charter member of Virginia Engine Company No. 1 in 1861, both served as chief of the old volunteer fire department, as well as its successor, the Virginia Paid Fire Department. Both were also members of the Virginia Exempt Firemen's Association.[18]

At least twenty men held the office of chief engineer of the Gold Hill Fire Department from its organization in 1867 until only one fire company in the department remained on active duty. Several of the chiefs served more than one term in office. W. D. C. Gibson was elected in 1867 and again in 1869. Alex McMartin served as chief in 1870, 1872, and 1874–1875. Abe Jones was chief in 1871 and 1873. John Marks had a dual term, 1883–1884, and Henry Baglin served 1889–1890 and again in 1893. However, the man who served longest as chief of the Gold Hill Fire Department was Michael Kennedy, a simple miner in Gold Hill who served 1876–1879, again in 1882, 1888, and finally in 1892, with a total of seven years' service as chief.[19]

While most of the elections went relatively smoothly, with few serious outbreaks of violence or campaign hijinks, there was a time when the taint of scandal put a damper on what normally would have been a jubilant victory celebration. In 1866, it was discovered that there had been twenty-three more votes cast than there were firemen eligible to vote for officers of the Virginia Fire Department. The election was immediately contested before the Board of Fire Delegates of the Virginia Fire Department. The meeting held to inquire into the apparent ballot box stuffing, Alf Doten wrote in his diary, was "animated," no doubt with considerable shouting and waving of fists. A week later the Board of Delegates again considered the election fraud. They declared M. R. "Riff" Williams, of Nevada Hook and Ladder Company No. 1, elected chief over rivals Jacob Young, Jr., and William Weir, both of Young America Engine Company No. 2. It was to be the last fireman's election tainted by fraud, although irregularities in other elections in Storey County continued into this century.[20]

Conducting the business of the Virginia Fire Department was left to the Board of Delegates, which was created by the Nevada Legislature in

1864, when it passed an act regulating the fire department. This act provided guidelines for the conduct of members and officers, and provided a general organizational plan for the efficient function of the department. The Board of Delegates was to be comprised of three members elected from each of the engine and hook and ladder companies, and two each from the hose companies. From the delegates were to be elected the president and other officers of the fire department, charged with overseeing the funds of the department, reviewing legislation affecting the department, and maintaining a liaison between the department and local government, while also helping settle disputes between the companies.

All fire companies were required to belong to the fire department. Membership in engine and hook and ladder companies was restricted to not more than eighty men, nor less than twenty-five; hose companies were required to have a minimum of fifteen members, and not more than twenty-five. When companies fell below the minimum number of members, there was a mandatory review before the Board of Delegates, with probable expulsion the result.

The chief engineer of the department was responsible for the construction of hose houses and cisterns, and the installation of fire hydrants. He had two assistants and was the supreme commander at any fire; all company foremen and assistant foremen were responsible to the chief and his assistants for the conduct of members, handling of equipment, and following of directions. The chief was responsible for investigating the causes of all fires and seeking punishment for persons criminally involved in fires or who vandalized fire equipment. He was also required to deliver a report on all fires and on the fitness of the fire department to the Virginia City Board of Aldermen.[21]

Firemen were exempt from jury duty and from serving in the militia or army, except in time of war, by mandate of the Nevada Legislature in 1864. The certificate of membership was all the proof required to claim the exemption. While most firemen took their duty seriously, there were many who joined the volunteer companies of the Comstock merely to be free of jury duty. The minutes of the fire companies and the columns of the local press reflect a distaste for such "dead wood," with calls from the press to oust the deadbeats and get on with the business of fighting fires.

The seriousness of this problem became apparent in 1865, when a bill that would have eliminated the exemption of firemen was introduced in the legislature. The members of the various fire companies united in anger to fight the measure. George Birdsall, delegate from Virginia

Engine Company No. 1, addressed a gathering of his comrades and other Comstock citizens and stated that "it had always been the perogative [*sic*] of firemen, in all well-regulated communities, that they should not be compelled to serve on a jury; that duty conflicting with their duty as good firemen and preservers of the public property." Birdsall considered the action of the legislature unparalleled in the Union, claiming that the firemen had rights that should be maintained. Considerable lobbying in the halls of the legislature, with the political pressure of more than three hundred members of the fire department, resulted in an amendment insuring the continued exemption of firemen from jury duty. In this episode, the fire department of Virginia City proved that, when united, it was a political force to be reckoned with.[22]

Later that year, the firemen again exerted themselves as a body, in a fight with the Board of Aldermen over proposed reductions in the salaries of the first and second assistant engineers from $1,000 to $300 each, and that of the secretary of the fire department from $1,200 to $300. Alf Doten wrote in his journals that there was "considerable trouble among the firemen" over the proposal, and that the department had promised to disband if the Board of Aldermen stood by its proposal. A protest march was organized on June 1, 1865, at three p.m. All of the firemen turned out with their engines and hose carts, meeting at Music Hall, where a series of resolutions were passed, including one declaring the fire department out of service. The firemen then reversed their helmets and returned in parade, with a brass band in the lead, to the various engine houses, where the apparatus was housed "tongue in." Later that night, when the fire alarm sounded, only Virginia Engine Company No. 1 rolled, and then with only a handful of firemen to put out the blaze.[23]

The following night another mass meeting was held at the courthouse to settle the problem. After considerable discussion about retrenchment in local government and the necessity for paying the officers of the fire department at their old salaries, the matter was amicably settled. The *Gold Hill News* happily reported:

> The Honorable Mayor and Board repealed the obnoxious ordinance reducing the salaries, with the understanding that private citizens should subscribe the deficient amount. The committee returned and reported favorably to the meeting. Another committee notified the Board of Foremen of the action. The boys at once put every machine in good working order and ready for duty. So ends the great broil between Virginia City officials and the Fire Department,

Yellow Jacket Hose No. 2

THERE WILL BE A REGULAR Monthly Meeting of the Members of this Company, at their Hall, TO-MORROW (Wednesday) EVENING, at half-past 7 o'clock.

A prompt attendance is required. The laws and by-laws will be fully enforced.

RICHARD MERCER, Foreman.

CHAS. BRODHUN, Secretary.

Gold Hill, May 10th, 1870. td

Attention, Liberty No. 1.

THE REGULAR MONTHLY Meeting of this Company will be held TO-MORROW (Wednesday) EVENING, May 11th, at the Engine House, at half-past 7 o'clock.

All the members are requested to be in attendance.

A. A. PUTNAN, Foreman.

E. L. STERN, Secretary.

Gold Hill, May 10, 1870. td.

Lincoln Hose Co. No. 3.

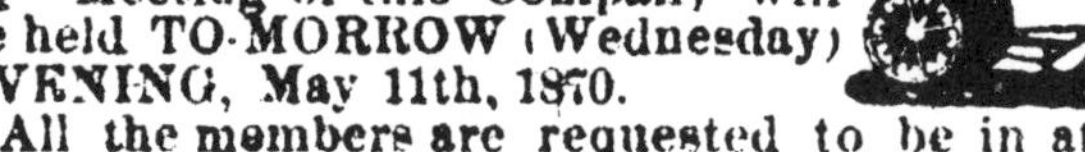

THE REGULAR MONTHLY Meeting of this Company will be held TO-MORROW (Wednesday) EVENING, May 11th, 1870.

All the members are requested to be in attendance.

H. C. BLANCHARD, Foreman.

ED. WRIGHT, Secretary.

Gold Hill, May 10, 1870. td

Fire company meetings often broke the monotony of a mining camp. They were announced in the local newspapers and were held in the engine houses. Some engine companies flew special flags from their flagpoles to announce meetings and rang their fire bells to call members together at the appointed hour. (Author's collection) and we believe that every person in that city slept sounder last night than they would with a breach between the two powers.[24]

The town of Gold Hill acted to prevent such a general strike when it passed an ordinance creating and regulating the Gold Hill Fire Department, which vested the corporate interests of the volunteer companies and their equipment in the town trustees. That move made the town virtual owner of the fire department and all of its equipment.[25]

Alf Doten: Editor In A Leather Helmet

SEVERAL OF the Comstock's most noted journalists joined the fire departments of Virginia City and Gold Hill. Joseph Goodman, proprietor of the *Territorial Enterprise*, was a member of Virginia Engine Company No. 1. Reporters Steve Gillis and F. C. Farrington were both members of Eagle Engine Company No. 3. In Gold Hill, Wells Drury, another reporter destined to be an editor, joined the ranks of Yellow Jacket Hose Company No. 2.[1] But none of the Comstock journalists stayed with their respective fire companies for more than a year except Alf Doten, editor of the *Gold Hill News*. When he was elected a member of Washoe Engine Company No. 4 in 1866, Doten wrote proudly in his journal, "Now, I am a regular 'fire jakey,' one of your 'be jesus fellers.' "[2]

Doten arrived in Virginia City after trying his hand at mining in the California gold fields and the Nevada mining camp of Como, high in the mountains overlooking the Dayton Valley. He was a native of Massachusetts, and in his journal he always noted "Founder's Day," a celebration from his childhood that usually meant a lively evening at a friend's cabin, or a cozy corner in his own cabin with a jug of the "Oh, be joyful." Doten corresponded regularly with his hometown newspaper, describing his experiences in the West. It was this writing experience that landed him a reporter's job with the *Gold Hill News*.

As a member of the Virginia Fire Department, Doten always turned out to "work the brakes" of the hand engine of Washoe Engine Company No. 4, a task not always easy to perform. In 1867 he wrote in his journal, "I went to my engine at cistern on B street & worked hard on the brakes for ½ an hour – we sucked cistern dry – Took up and housed machine –

308
Gold Hill News editor Alf Doten was extremely proud of his membership in Washoe Engine Company No. 4. (Doten Collection, Special Collections, University of Nevada, Reno, Library)

Staid & answered at roll call – So did Brier – While at the fire I fell & cut my right hand inside, next wrist – bled considerably."[3]

Doten was especially supportive of the fire companies of Virginia City and Gold Hill in his role as reporter, and later editor, for the *Gold Hill News*. His accounts of the elections, drills, meetings, social events, and other news of the fire departments were detailed and frequent, providing the latest information on what was, at the time, one of the most powerful organizations on the Comstock. When times were bad for Comstock firemen, such as the days following the Great Fire of 1875, when other newspapers were clamoring for the disbanding of the volunteers, and blaming them for staggering losses, Doten was the only newspaperman to express support of the old system.

Through all the good and bad times of the fire department, Doten remained proud of his membership in Washoe Engine Company No. 4. He had a photograph taken of himself dressed in the uniform of "one of 4's boys," and sent copies to friends and relatives. His speech at the housing of Knickerbocker Engine Company No. 5, in 1867, probably best illustrates his feelings for the Virginia Fire Department:

> Gentlemen of the Virginia Fire Department, and No. 5 especially, I never made a speech in my life, and never expect to. But were I a Webster or Clay, I would make you a speech that would make your hair stand on end. Were I a Demosthenes or a Cicero, or any other man, I would go on and tell you how proud I am of the Virginia Fire Department, but as it is, all I can say is, in the language of Old Winnemucca: "Bully Boys." I ain't much of a fire jakey, but I tell you boys when I turned out last "Fourth" in the procession with a leather hat on my head, and my emaciated form done up in a sanguinary shirt, I felt prouder than I ever did since God Almighty made me.[4]

It was a sad time for Doten when the old volunteer companies of the Virginia Fire Department disbanded and the fire department was dismantled. He mournfully noted in his journal each payment of his share of the treasury, as the property of Washoe Engine Company No. 4 was sold off.[5] But Doten was not through with his old comrades of the volunteer fire department. As he proudly recounted in 1878, "I was elected a member of the Exempt Fireman's Association at Virginia, at their regular meeting on Sunday evening – proposed by L. Kaplan, formerly of Washoe Engine Co. No. 4 of which I also was formerly a member – Both of us staid with the Co. until disbanded and displaced by the present paid Fire Dept."[6]

Once again he was associated with the men who had worked the brakes of No. 4 and the other "masheens" at fires on the Comstock. His affiliation with the Exempts was a long one. He had the distinction of being the only Comstock journalist among those who were members of the Virginia Fire Department, ever to become an Exempt. His journalistic talents were called upon many times by that group. He composed resolutions of thanks for donations to the association, and wrote resolutions mourning the passing of members, who were often escorted by their comrades to the Exempt Firemen's Cemetery.[7]

Doten remained a member of the Exempts for many years. In 1897, he was warned that he was delinquent in dues to the amount of $9.00, and that if they were not paid, he would be "read out of the association." His old colleagues wanted him to remain a member, but his fortunes had slipped. In his journal of May 8, 1897, Doten noted that notice had been sent by Secretary Fred Plunkett, who wrote, "Alf, I am instructed to forward this by the Association as the last appeal. Plunk." It was to be one of Doten's last recorded references to his old friends or to the association they had formed to take the place of the old volunteer fire department.

When Alf Doten died in 1903, there were hardly enough of the old "boys" left to turn out for his funeral in Reno. The man who had served as trustee, secretary, and president of the Exempts, and who had written so many resolutions of mourning for his comrades, was not even accorded the customary final honors that had been given to so many others who had worn the fireman's red shirt and leather helmet on the Comstock Lode.

II. Dangerous Work

"The firemen of this section never stand back when a big fire is destroying things," the *Gold Hill News* bragged in 1869, "but they rush in and put it out, even if they have to roll in it to do so."[1]

There is a certain foolhardy bravery that seems to exhibit itself when volunteer firemen pit themselves against the flames. The firemen of the Comstock had the scars and often wore the bandages to prove it. Lending credibility to its boast, the *Gold Hill News* told its readers, "Owing to the frequent fires at Virginia of late, there are more crippled or damaged firemen than we ever saw there before. We know of several who go limping about from having been run over by machines or hose carriages, to say nothing of numerous ambitious ones who have their faces or hands badly blistered."[2]

The men in leather helmets wore their cuts, bruises, and bandages proudly. They knew the dangers of facing a flaming building, but it didn't matter. They were firemen, heroes to the town, and expected to take chances. All were tried and true men—brothers in the battle against fire. The dangers were clear, and there were some who seemed to seek them out at every opportunity.

Riff Williams was one such fireman. The former chief of the Virginia Fire Department jumped into action in 1868, when the blacksmith shop at the Yellow Jacket Mine in Gold Hill caught fire. Williams took the hose from the Yellow Jacket works by himself and stood between the main hoisting works and the inferno of the burning blacksmith shop, keeping the flames away from the important hoist house. It got so hot where Williams made his stand, that he finally had to wrap himself in a heavy woolen blanket, which was kept wet by another hose throwing water upon him from a safer and more comfortable distance. The *Gold Hill News* reported:

> We have seen our brave little salamander, Riff, in just such positions before, on several trying occasions at Virginia, when even the coat was burned from his back but he never flinched as long as he could see Tom Peasley, Dick Paddock, Ben Ballou or any of the other boys about. Riff is as much in his element at a fire as a Scotch terrier in a rat pit.

Williams's hair was singed, but he managed to keep the fire from spreading without getting hurt. Former Gold Hill Town Marshal W. I. Cummings was not as fortunate, however. When the hose he was working suddenly jumped, he was thrown from a pile of old lagging about eight feet high, down upon a pile of old iron castings. Even with a dislocated right ankle and several bruises and sprains, the plucky fireman continued to play water upon the blaze until it was out before seeking treatment for his injuries.[3]

Sometimes the elements worked against the stalwart efforts of the fire laddies. Dust, hot ashes and cinders, and rocks blown about by the wind were a constant hazard for the fire companies. In the winter months, cold weather had a very dramatic effect. When firemen rallied in 1869, to fight a fire in a brick building on C Street, the weather turned out to be the most formidable opponent, as wet clothes froze on the firemen. "We have seldom seen a worse night for firemen, as far as winter weather is concerned, and all who participated on that occasion are worthy of the highest commendation," the local press reported.[4]

When Jule Clemen's brothel caught fire in 1865, however, it was the firemen who created their own hazard, with a frontal assault on the blaze. They climbed up on the porch and hung from the balustrade, almost upside down, getting the first stream upon the fire. "Flames danced about and licked their faces & hands, but they were soon masters of the situation," Alf Doten reported.[5]

Hidden dangers were always present in the fires of Virginia City and Gold Hill. There might be one hundred pounds or more of Giant Powder stored in the back room of a burning house, unbeknownst to responding firemen. There might be chemicals stored in an assay office, emitting deadly fumes in the heat of the blaze. Cans of paint could explode on the shelves of a paint shop with the same effect as a cannon shot. No matter what the dangers, it was never an easy task to fight the fires in the mining towns, and there was always the danger that someone might be trapped or killed.

When the Gould Mill caught fire in 1864, several of the firemen were "badly salivated" by inhaling the fumes of quicksilver from the amalgam in the pans.[6] When the paint shop of Happs & Finney, just north of the Wells Fargo office, became engulfed in flames in 1865, some firemen were injured by falling timbers from the burning awning, which dropped several of them to the ground below. The *Gold Hill News* reported, "Barry was taken, insensible, to a store across the street, where Dr. Bryarly examined him, and found him severely but not fatally injured, he having suffered a blow on his head, which fortunately did not constitute a fracture. Wright was stunned by the fall, and somewhat scratched."[7]

It was a miserable night for firemen fighting the blaze set by an arsonist in the St. Louis Brewery, on C Street in Virginia City, in 1868. Several of the "fire laddies" were overcome by dense and rank coal oil smoke, but the biggest danger was the imminent collapse of the roof and brick walls as the firemen rushed about, working the brakes of the hand engines and taking hose inside the building to try to stop the flames. Although the evening was barely tolerable, the firemen broke out in song several times as they pumped their hand engines in the cold, disagreeable, drizzling rain, which helped put the fire out.[8]

Fighting fires was dangerous business. Getting there, however, could be just as dangerous. The race to the fire was important to the volunteer firemen of the Comstock. There was fierce competition for the honor of having put "first water" on any fire, which resulted in some spectacular and dangerous races along the town's rough dirt streets.

When buildings on the corner of C and Mill streets caught fire on September 9, 1869, the firemen put heart and soul into the run to the blaze with their engines and hose carriages. The first building was blazing fiercest, and the buildings opposite the fire had begun to smoke from the radiated heat when Young America Engine Company No. 2 came dashing down the street and almost through the flames. The faces of the men on the engine were scorched from the intense heat, and the paint on the right side of the engine was blistered and destroyed. Right on the heels of No. 2 was the jumper of Rooster Hose Company No. 1, attached to Virginia Engine Company No. 1, which knocked down a Mr. Breed, owner of the Garden Valley House located in the burning district. Before he could get up and recover from his first encounter, the jumper of No. 2 came rushing through, running over him. Had it not been for a dozen men who rushed to his assistance and dragged him out of the way of the other oncoming equipment, Breed would have been killed by a third collision, if not roasted alive as he lay within twenty feet of a blazing inferno.[9]

Such mad dashes prompted local newspaper editors to warn firemen and citizens alike to take caution during the alarm of fire, "for a single act of carelessness might cause the loss of several lives." This particular warning was issued in 1864, after several firemen were injured while running to a fire at the quarters of the provost guard. According to the report in the local newspapers:

> A fireman named Charles Safford was thrown from the tongue of the "truck" on the corner of C street and Sutton avenue, and striking the side of a building suffered a fracture of the leg. He was taken to rooms where a physician attended to the injury, who pronounced it very painful though not of a dangerous character. Another fireman, named Mike McGowan, while running with Knickerbocker Engine No. 5, was run over by the engine, on Union street, and his shoulder dislocated, and it is feared some internal injuries sustained.[10]

On another occasion, in 1864, Eagle Engine No. 3 ran off a fifteen-foot embankment near the Central Mill while running to a fire in the northeastern part of the city. Fortunately no one was injured, and the engine sustained only a few scratches and dents.[11] Virginia Engine No. 1 was not as lucky in running to another fire, however, in 1865. When a few firemen reached the engine house at the first sound of the alarm, they wheeled the engine out and began the run, confident others would soon join them. But the engine broke away from the firemen and piled into a stone house, smashing the tongue, wheels, and other gear, "in a decidedly unprofitable manner." Bruce Garvey and Billy Nolan, who had been on the tongue, narrowly escaped with their lives, and sustained severe bruises. Tim Cronin was not as fortunate, and suffered a broken ankle in the impact. Several bystanders were also injured as the ponderous engine whirled down the hill in the darkness. "It seems wonderful that more people were not injured or killed outright," commented Alf Doten.[12]

When the hose carriage of Liberty Hose Company No. 1 ran out of control on the steep streets of Gold Hill in 1867, there were similar injuries. Jules Hamburger was run over by the wheel, which broke his leg

Abe Jones in 1870, chief of the Gold Hill Fire Department. (Comstock Firemen's Museum)

halfway between the knee and ankle. He was carried to the Vesey House, where he received the attentions of both a doctor and a generous bartender.[13]

Arsonists caused an alarm in 1867, when they set fire to the Belvedere Hotel. It was an exciting, if somewhat dangerous, run for the volunteers of Young America Engine Company No. 2. Rounding the corner at Taylor and D streets, the engine careened into a small frame building, tearing out a whole side and the end of the structure and severely damaging the engine, not to mention several firemen. James Phelan and John Leconey, who were on the tongue, were bounced around and received several bruises. David Stearns, who was on the brake, was thrown against the building while going down Taylor before reaching D, and received a severe cut over one eye.[14]

Manning the engines could also be hazardous to a fireman. Exhaustion came easily after only a few minutes of working the brakes (i.e., pumping water).* Some firemen were knocked down by the down-stroke of the brakes. On one occasion, a fireman of Eagle Engine No. 3 lost several fingers when his hand was pinched in the down-stroke of the brakes.[15]

All the engines in Gold Hill and Virginia City responded to the arson fire at the Ashland House in 1870. Before the fire was beaten down, however, one fireman lost his life through an unusual accident—drowning. "Saddest and most serious point of all," the *Gold Hill News* commented, "was the death of Martin Panian, a member of Liberty Engine Company, by falling into a cistern, where he drowned." Panian was proprietor of the Nevada Brewery Saloon in Gold Hill, and was a well-known fireman in that town. Though he had run to the fire with his own company, circumstances soon found him working the brakes of the hand engine of Monumental Engine Company No. 6, near the open hatch of a cistern. He apparently stepped into the black hole and fell into the water below. "He made no noise, and so quickly was he out of sight that there was some dispute as to whether he really fell into the cistern or not. In the great confusion consequent upon the occasion, it was several minutes before effective measures were taken for his rescue," the

*The author has worked the brakes of hand engines from Virginia City, as well as Benecia, Columbia, and San Francisco, California, and can attest to the effort that must be expended to get water into the hose and then keep a steady stream on a fire.

newspaper said. A fireman lowered himself into the twelve-foot-deep cistern and felt around, but could not locate the body.

Even with the aid of a lantern lowered into the inky darkness, Panian's body could not be located:

> A ladder was next procured and placed into the cistern, reaching the bottom. Owen Cook, foreman of the Imperial-Empire shaft, now passed down this ladder, and diving beneath the water, searched every corner of the cistern. He had to dive four times in order to do this, but the fourth time he succeeded in finding the body lying upon the bottom, and dragged it with him up the ladder. The unfortunate victim had, however, been now in the water some twenty minutes, and all efforts at resuscitation proved ineffectual, although continued for over an hour. The body was taken to Gold Hill and properly laid out in the hall of Liberty Engine Company.

A coroner's inquest the next morning determined that the thirty-three-year-old Austrian native had fallen into the cistern, striking his nose, and that he had drowned in at least eight feet of water.[16]

While accidents to the volunteer firemen were common in the normal routine of running to the fires and working the hoses and engines, there were some who sought out danger and teased death in the face of disaster. disaster. Among them were the firemen who fought the Yellow Jacket Mine fire of 1872. They surely remembered the thirty-seven men who perished from the flames, foul air, and smoke on a similar occasion, "Black Friday" in 1869. It was no wonder then, that when the mine caught fire again in 1872, there was a general and particularly hasty exodus from that shaft, as well as the Kentuck, Crown Point, and Belcher shafts, all of which were connected.

"The horrors of that great calamity [1869] of which we speak are still too fresh in the minds of all miners to allow them to stay for a moment where there is a smell of smoke or an alarm of fire given," explained the *Gold Hill News*. But the draft forcing the fire and smoke through the mines that day in 1872 was different from the deadly blast of air which had killed so many in 1869, and as a result there were no deaths in 1872. Still, the fire had to be put out before it destroyed three of the richest mines in Gold Hill. Sandy McMartin, former chief of the Gold Hill Fire Department, who was employed as the foreman of the Crown Point Ravine Water Works, made preparations to send down a stream of water through hose by way of the Crown Point shaft. Instead, six members of

Liberty Engine Company No. 1 stepped forward to volunteer to go into the inferno below ground. Hi Holman, foreman of the engine company, and Robert Beegan, foreman of the hose company, led the team of David Humphrey, Jerry Sheehan, John Marks, and James Moore down the Belcher shaft and into the Crown Point and Kentuck mines. Finally, the team made its way into the Yellow Jacket shaft and directly to the fire. With the assistance of some miners who had gone into the shaft from another direction, they worked to extinguish the fire with a few buckets of water. "The boys worked pretty lively for a short time, getting a supply of water from a barrel nearby," the newspaper reported. "They are deserving of much credit for their promptness and efficient strategy."[17]

Time and again, the men in leather helmets placed their lives in danger. They struggled with hose in the face of burning walls which might tumble down on them at any time; they worked their way through smoke and deadly fumes to help rescue fire victims and prevent the spread of the "fire fiend" through their town. Often they were injured. Sometimes they were killed. But danger never deterred them, because they were proud to be firemen—heroes in their leather helmets.

The Death of "Old Butt"

THERE WERE few fatalities as a result of the fires on the Comstock Lode, but when someone did perish in a fire, there was genuine sadness. This was certainly the case for the tragic death of one Comstock resident during a fire on September 19, 1871. The event is best chronicled in the pages of the *Gold Hill News* of October 14, 1871:

> Most everybody at Virginia knew or has heard of Pat Mulcahy's old dog "Butt." He was a good-sized full-blooded bulldog, and one of the most sedate and sensible dogs in the world. He knew more than some members of the Nevada Legislature, and he certainly was more incorruptible. When Mulcahy was sheriff of this county, Old Butt felt the dignity of the office fully as much or more than his master, and he kept as good an eye on the prisoners in the County Jail and things generally about the office, as though he had been

Undersheriff or Deputy. And Old Butt was the hero of many a well-fought battle, always coming out chief. He never would fight a small dog, he was too high-toned for that, but always went after any obnoxious dog of his size or even bigger, Butt wasn't particular. Yet he wasn't a quarrelsome dog. He didn't go about spoiling for a muss, like some bulldogs, but he would stand no nonsense from any of them. If any disagreeable dog crowded him or acted as though he wanted anything, Butt was the very dog who could give it to him. Old Butt was a resident of Virginia for nine years, and was brought over from Nevada City by his master when he was a yearling pup. He ran with the Fire Department, and was quite effacious in various capacities, being a good citizen, with kindly regard for children and ladies, never biting anybody, yet letting everybody know that they must do right and not fool with anything he was guarding or looking out for. Last of all, he was at Major Ferrend's livery stable, on C Street, where his master was superintendent, having Old Butt for second boss. When the great fire occurred on the 19th of last month, the stable was included among the buildings destroyed. Old Butt was running about in a considerable state of excitement, and some friend, fearing that he might run into the burning stable and be destroyed, put a rope about his neck and tied him carefully in the entry of Dias & Glauber's building, at the foot of the stairs, next to Kelly & Lowenstein's store. The building being fireproof, this friend doubtless thought Old Butt would be safe, but alas, the iron doors became red hot and the woodwork inside all around the unfortunate dog was burned, and the next day he was found dead, with his nose close to the iron doors. Poor Old Butt. His disconsolate master never wants to own another dog.

III. Gold Hill's Great Fire of 1870

Throughout the history of the Comstock Lode there have been numerous instances of heroism by the firemen of the many engine, hose, and hook and ladder companies. Men have seemingly disappeared into a sheet of flame, only to emerge from the smoke and confusion carrying someone to safety. Others have managed to position themselves between burning structures to more effectively fight the fire, working hose lines and nozzles until the blaze was stopped in its tracks, but suffering serious burns or other injuries in the process.

Becoming a fireman meant having a certain amount of courage and being ready to demonstrate that courage when the occasion arose. That occasion arose for many Gold Hill firemen on July 5, 1870, when more than forty buildings in the heart of the community were destroyed in Gold Hill's greatest fire.

Although the summer of 1870 had been extremely dry, the townspeople were not greatly concerned about the possibility of fire, due to an abundance of water stored in the town reservoir, which, in turn, fed the new system of fire hydrants lining Main Street. The fine equipment and men of the Gold Hill Fire Department added even more assurance of protection, but when the alarm was sounded shortly before two a.m., the hose companies attaching their leather hoses to the new fire hydrants found a dry system. The night watchman at the reservoir had fallen asleep and did not hear the alarm, so there was no one ready to open the gates, allowing the town's hydrant system to fill with badly needed water. This gave the fire a tremendous head start, as it moved from

The main business district of Gold Hill, from Fort Homestead. (Special Collections, University of Nevada, Reno, Library)

Coiture's Grocery and Variety Store, where it had started, to adjacent buildings, and then quickly leaped across Main Street to the buildings on the opposite side.

Each of the fire bells on the Comstock took up the alarm chorus sounded in Gold Hill. Before long, the engines of Monumental Engine Company No. 6, Knickerbocker Engine Company No. 5 and Young America Engine Company No. 2 were clattering along the streets on their way to Gold Hill from Virginia City, followed by four men on the hose carriage of

Eagle Engine Company No. 3. Every possible effort was made by the firemen of Virginia City to assist their brethren in Gold Hill.

Gold Hill Fire Chief Sandy McMartin sent a man to the town reservoir to open the gates and direct water into the hydrant system, but the man became so excited in trying to turn the valve that he broke his wrench. Only after several minutes of extreme exertion was he finally successful in opening the water system. Father Clarke, although not a fireman, was busy recruiting men to work the brakes of the hand pumpers and man the hose as the fire began to spread throughout the town. Both sides of Main Street were on fire, which created a tunnel effect over the center of the street, where the flames arched to meet.

A. A. Putnam, foreman of Liberty Engine Company No. 1, called for volunteers to help haul the hand engine of the company to the lower end of Main Street to attempt to stop the spread of the flames into lower Gold Hill. Ed Reese, Patrick Kelley, Heber Holman, Billy Rower, Sheriff Cummings, James Moore, a man named Peabody, and a volunteer from Silver City whose name was never recorded, stepped forward, grasped the tongue and ropes of the engine, and immediately started on the run through the tunnel of hell. "They made the trip in safety," the *Gold Hill News* reported, "but every hero of them got terrible scorched about their hands and face. Putnam and Kelley especially their faces and necks burned to a blister." But the run had been worth the pain, as the small group of heroic volunteers maneuvered the heavy engine down the street and through the flames, getting into position and put "first water" on the fire from that section of town. The brave crew was successful in its mission, preventing the spread of the inferno to more homes, businesses, and mine buildings. Playing water upon spot fires and cooling smoking wooden structures, the firemen were able to halt the blaze just below the Vesey House, which had miraculously survived the onslaught of the inferno.

The fire had spread along each side of Main Street for the entire width of the town, between the railroad on the west side, and High Street and Fort Homestead hill on the east. At the lower end of the fire, destruction had extended to the house of John Lambert, which had its paint scorched off the siding. At the upper end, the flames had moved through an alley to the rear offices of the *Gold Hill News*, sweeping away the buildings on each side and in the rear as well as the awning and sidewalk of the newspaper. The *Gold Hill News* was spared destruction, with only smoke damage and some scorched woodwork around doors and windows in the "fireproof" brick structure. Thankful for being spared by the fire, the newspaper noted that its iron doors were "of white heat for some time. 'All's well that ends well.' We are thankful as we sit in our shirt sleeves writing, amid the heat, smoke and embers of the surrounding ruins. The senior editor's residence is demolished, but the *News* offices still existeth."

At the height of the fire, war seemed to break out, as a lively bombardment of rockets, firecrackers, and petards rained down on the volunteers from the building in which the fire started. Firemen working hoses had to retreat from time to time as the cannonade centered on them. Even as the ashes cooled, there were still echoes of firecrackers making their shot-like noise in the ruins.

More than $150,000 in losses were counted, for which there was only an estimated $80,000 of insurance coverage. The major buildings destroyed included Bittner's Hardware Store, J. Jones's apothecary shop, the Washoe Saloon, Philadelphia Brewery, Weigand assay office, San Francisco Restaurant, and the parsonage of the Episcopal church. Also standing in ruins was the engine house of Yellow Jacket Hose Company No. 2. Records, flags, decorations, and tools were lost in the blaze while firemen fought to save other buildings in the town. In all, there were at least forty major structures destroyed, as well as a hundred smaller sheds, outhouses, barns and other structures not accounted for in the report of the fire. Nearly all of the buildings destroyed had been made of wood. The hardware store of Bittner & Company was built of brick, with large iron doors and shutters, making it fireproof by standards of the time. But the roof, constructed simply of tin on wood, proved to be the vulnerable point in the structure. Weigand's assay office and the Gold Hill Market had both also been thought fireproof before the blaze.[1]

Despite the wholesale destruction of the center of the Gold Hill business district, along with many homes, the residents of the town began to rebuild with hopes of new prosperity. However, almost one year later, in June of 1871, Gold Hill was again devastated by fire, only hours after the *Gold Hill News* published an extensive report on the fine condition of the volunteer companies comprising the Gold Hill Fire Department.

The blaze was caused by an arsonist setting fire to an unoccupied wooden building near the Eclipse Mill. The alarm was sounded by the fire bells and steam whistles in Gold Hill, and was relayed to Virginia City by the bell atop Monumental Engine Company No. 6 on the Divide. Within a short time, the Gold Hill firemen were joined by the volunteers manning Monumental Engine No. 6 and Knickerbocker Engine No. 5,

Gold Hill, the day after the great fire of 1870. The heavy hand pumper of Liberty Engine Company No. 1 was hauled in a heroic dash through the flames, with the result that the town's lower business district was saved. (Special Collections, University of Nevada, Reno, Library)

along with the hose carts of Young America Engine Company No. 2 and Virginia Engine Company No. 1. Also on hand in this early display of mutual aid were Chief James Malone, Second Assistant Chief J. P. Bell, former Chief Peter Larkin, Washoe Engine Company No. 4 Foreman James N. Cartter, and many other members of the Virginia Fire Department. Through their quick efforts, the fire was limited to one side of the street, and burned mostly small stores and a few houses. Two of the largest structures destroyed were the Eclipse and Granite mills, a loss totalling $33,400, with insurance between the two amounting to only $5,300.

The *Gold Hill News* reported that a man sleeping next door to the O'Neil house, where the fire originated, heard one or more persons walking about in the house shortly before the fire, giving support to the suspicion that an arsonist had been at work. "What a pity that the villain escaped from it," the newspaper stated. "Half an hour's roasting would have been the very thing for him." There were no serious injuries during the fire, and no one made dashes through the flames with engines or hose carts; but the newspaper called every fireman there a hero and gave special thanks for the assistance rendered by the various companies and members of the Virginia Fire Department.[2]

The time would come when the firemen of Gold Hill would return the favor to Virginia City, although with less success than had been achieved by their neighbors for them.

IV. The Social Life of a Fireman

While there was a great deal of responsibility to being a member of the Comstock volunteer fire companies, there were also lighter moments of fun and "jollification." It was a toss-up as to whether the company's annual fancy dress ball or the Fourth of July parade was the favorite event of a fireman. Both events offered a chance to publicly demonstrate pride in his company and dress up in his red shirt and leather helmet.

Some of the most elaborate balls in the mining camps were given by the fire companies, and there was never a good excuse for missing one of these formal affairs, as fireman and newspaper editor Alf Doten noted in his diary on January 1, 1867:

> Oro Hotel, kept by Mrs. Wylie, on fire in the upper stories—3 stories high, wooden building next to the Union office on D street—Engines soon got 2 or 3 streams on & put it out before it could catch any of the adjoining buildings which were all of wood . . . I went up to my engine [No. 4] at cistern on B street & worked hard on the brakes . . . Then went to Piper's—Then to No. 4's Ball at Athletic Hall—crowded—probably 80 couples present—Best ball of the season.[1]

The firemen spared no effort or expense in preparing for their annual balls. Some of those events were designed to raise funds for the company. Others were "invitation only" affairs for the members and close friends of the company. In the course of one year there might be as many as ten fancy dress firemen's balls on the Comstock—one for each of the engine companies, the hook and ladder company, and the various hose companies of the fire department. Some were scheduled to coincide with holiday periods, while others were scheduled on the organization date of the volunteer company, making it a grand anniversary ball.

The ball of Liberty Engine Company No. 1 in Gold Hill, on February 24, 1872, included a full dinner served next door to the dance at the engine house. Both the dining room, in Miners' Union Hall, and the engine house were "decorated handsomer than ever before, with flags, pictures, emblems, etc., and one splendid arch of flags spanned the center of the hall," the *Gold Hill News* told its readers. Because the weather had left the streets muddy, Liberty Engine Company went to the expense of chartering Jake White's horse-drawn bus, which made regular runs from Virginia City locations to the dinner and dance, with the promise of a return trip following the affair.[2]

The ball of Young America Engine Company No. 2, held at Athletic Hall in Virginia City on March 21, 1870, featured more than one hundred cages of singing canaries suspended throughout the hall. There were also some parrots, a mockingbird, and four gamecocks in cages, "all of which contributed their quota to the general music occasionally," commented the *Gold Hill News*.

Young America was noted for outstanding firemen's balls and is credited with holding one of the most elaborate, on March 12, 1873, in the Odd Fellows' Hall in Virginia City. Professor Carl York led a band of sixteen musicians, who played the most popular songs of the day. Although the music was pleasing, it was the decoration of the hall that brought praise from the local press. The decorating had been supervised by Joe Josephs, and included garden walks, flower beds, trellised arbors, evergreen trees with singing birds in the branches, and a mischievous gray monkey. An old gray goose wandered about the hall, occasionally eyeing the fountain in the center of the hall as a prospective bathing pond. The pond held goldfish and frogs, which were sprinkled by jets of water the size of knitting needles, thrown to a height of four or five feet. A second basin caught the overflow and the entire centerpiece was surrounded by a small trellised fence. Dancing continued until dawn's light appeared through the windows of the hall. The newspapers praised the ball enthusiastically, noting that it was the first ever held in the new Odd Fellows' Hall.[3]

Another favorite annual occasion for those sporting the red shirts and leather helmets was the Fourth of July. Then hand engines and hose carriages would be handsomely decorated with ribbons, flags, flowers,

GRAND

ANNUAL UNIFORM BALL

.........OF.........

LIBERTY HOSE CO. NO. 1

....TO BE HELD AT....

MINERS' UNION HALL, GOLD HILL

......ON......

Friday Evening......June 5, 1891

Music by Silver State Brass Band

TICKETS..................................$1 00
(Admitting Gentleman and Ladies.)

The fire companies relied heavily on the newspapers to help promote special events such as the annual uniform ball. Complimentary tickets were often sent to editors who would review the event in glowing terms in the newspaper the next day. (Author's collection)

FIRST GRAND

Uniform Ball

To Be Given By

Divide Hose Co.

(NO 2,)

......A T....

NATIONAL GUARD HALL

.... ON ....

Friday Ev'ng, April 7, '99.

Music By Full National Guard Orchestra.

TICKETS..................................$1 00
(Admitting Gentleman and Ladies.

evergreens, and other items, to be pulled down the rough dirt streets by the entire company, all dressed in their best firemen's outfits. Such occasions warranted entire columns in the local newspapers, describing the holiday procession in detail and giving special attention to the entries of the various fire companies. One such report appeared in the *Gold Hill News* on July 5, 1871:

> Virginia Engine Co. No. 1 had their machine handsomely trimmed with flags, wreaths, etc., and drawn by six gray horses, the little Goddess of Liberty riding beneath the beautiful canopy on the machine, was Lizzie Denning. Rooster Hose Co., attached to this company, turned out in good style, with their hose carriage beautifully trimmed, and with a live fox and a game rooster mounted upon it. Young America Engine Co. No. 2, made the finest display. Their large machine was splendidly decorated and drawn by eight horses, with four colored grooms riding the near horses, dressed in red jackets and caps, white pants, etc., and Gussie Scott, with three attendant maidens, rode as the Goddess of Liberty, under the pretty canopy of the machine. Good Will Hose, attached to this company, composed of young men, followed with their hose carriage elegantly trimmed, and were succeeded by fifty little boys dressed in black pants, white shirts, and straw hats with red ribbons bearing 'Young America' inscribed thereon in gilt letters. They drew a huge allegorical picture or transparency, representing Peace and War, with the portraits of Washington, Lincoln and Grant, and they were about as proud and happy a set for their inches as anybody ever saw. Eagle Engine Company No. 3, made an excellent showing. Their fine machine was very tastefully trimmed with evergreen festoons, flowers, flags; and Isabella Alchorn, who rode as Goddess of Liberty, beneath a very handsome canopy, was about as pretty and tastefully arrayed a little divinity as could have been desired. Washoe Engine Company No. 4, had their most effective and neat machine trimmed and decorated in a very tasteful manner, surmounted with a splendid canopy, and Miss Carrie Clark rode as the Goddess of Liberty. She was very appropriately dressed, and the prettiest Goddess we ever saw on any similar occasion. Knickerbocker Engine Company No. 5, did not have any goddess or canopy on their machine. They turned out the greatest number of firemen, however, and their machine looked just right in the eyes of all good judges of fire engines. It was a pattern of neatness and efficiency. One of the finest features in the procession was Neptune Hose Company, attached to the Knickerbockers. It is composed of some of the best and smartest young

This 1863 view is one of the earliest known photographs of Virginia City. It shows the members of Young America Engine Company No. 2 decked out in their finery for a Fourth of July parade. In the center is the company's double-end-stroke Rogers hand pumper. (Special Collections, University of Nevada, Reno, Library)

> men of the city, and their hose carriage, recently from Sacramento, was tastefully trimmed, and showed to excellent advantage. The members were dressed in black pants and white shirts, and blue caps with red bands. Mounted beneath a beautiful canopy, in front of the hose reel, sat little Jimmy Fair, son of J. G. Fair, the well known mine and mill owner. Jimmy personated Neptune, or rather son of Neptune, and was dressed in blue pants and jacket and broad collar trimmed with white stripes, usual to the near sailor costume, and a blue ribbon on his jaunty little straw hat bore the word Neptune in gilt letters. He was seated in the stern sheets of a miniature boat, with a little steering oar, or paddle in his hand, and looked as dignified as the captain of a revenue cutter, or a young Admiral. Monumental Engine Company No. 6 had their machine also appropriately decorated, and had two little girls riding beneath a pretty canopy. Their hose company of young men were tastefully dressed and all presented a fine appearance.[4]

Any occasion to don their bright red shirts, fancy belts, and leather helmets was a special one for the firemen of Virginia City and Gold Hill. So in addition to their annual balls and parades, they often created special events which enabled them to dress up and demonstrate their pride in being volunteer firemen, and in their respective companies. One such special occasion, a picnic sponsored by one company but attended by all, was held in the carpenter shop building at the mouth of the lower Gould & Curry tunnel in 1868. The party lasted for three days, commencing on the Fourth of July, and included such events as pick chases, foot races, sack and wheelbarrow races, and climbing greased poles. In the evenings there were dance contests, with prizes for the best pair of waltzers.[5]

The housing of new equipment, or of old equipment in new quarters, called for a grand housewarming, with members attending in uniform. One of these events was held in 1870, upon completion of a new engine house for Young America Engine Company No. 2, on C Street. Two large tables of food extended the entire length of the hall, with "rivers of lager and oceans of punch" available for those in attendance. All of the officers of the fire department attended, and James Phelan, first assistant engineer and a member of the company, served as presiding officer. Also in attendance were the mayor and board of aldermen, together with large delegations from the other companies, and a sizeable gathering of townspeople. Several speeches were made, followed by impromptu songs from members of the audience, including Judge Livingston, who favored the company with a song in his native Scottish dialect.[6]

The holiday season also prompted a dazzling display of uniforms at the various engines houses, as the companies spread holiday cheer among their friends and the townspeople. In 1869, for instance, Liberty Engine Company No. 1 of Gold Hill opened its engine house on Christmas morning and offered a cup of eggnog and traditional Christmas songs in front of the company's hand engine.[7]

Although not normally as formal as the fancy dress balls or parades, benefits for the various fire companies were equally gay, and often called for members to don their regalia. These benefits were sponsored by the owner of the theater or opera house in which they were held, or might even be offered and sponsored by the theatrical troupe or circus appearing in town at the time. They were held in the various opera houses of Virginia City and the Theatre Hall in Gold Hill for as long as the volunteer companies were in existence, and later, to benefit the Virginia Exempt Firemen's Association. Whether it was Shakespeare, an opera, or a horse comedy, the performances were always well attended and appreciated by the members of the fire department.

In 1864, a special pavilion was constructed for a benefit performance by Wilson & Zoyara's Circus. The *Gold Hill News* reported on August 18 that the performance was "crowded to its utmost capacity, it appearing as though everybody in town was there with wives and family. Huge delegations of firemen from Virginia City attended the event, some in uniform, with the largest delegations representing Young America Engine Co. No. 2 and Eagle Engine Co. No. 3." A special benefit festival was organized by the ladies of Gold Hill in 1869, to help construct a new engine house for Liberty Engine Company No. 1. Theatre Hall was decorated with evergreen boughs and firemen's paraphernalia for the occasion, and the "four bits admission" included three hours of dancing and the chance to partake of strawberries, ice cream, and lemonade. There were also "fancy articles for sale," including silverware and other items, to help raise the needed funds.[8]

Although a sad occasion, the funeral of a member of the fire department was no less a formal social event. It called for all members to turn out in dress uniforms and march in mourning of their fallen comrade, even though he might have belonged to a rival company. Often, the engine or hose carriage of the company would be draped in black, as would the engine house, from whose bell would toll a steady peal.

Services were occasionally held in the engine houses themselves, or the company would rent a large hall for the funeral. Following testimonials and eulogies for their fallen friend, the firemen would escort the body to the fire department's cemetery or to a fraternal society cemetery, where graveside services would then be conducted.

For fun and excitement, though, nothing could match the firemen's foot races, musters, or tournaments. One of the most exciting and controversial tournaments was sponsored by Young America Engine Company No. 2 on C Street, in 1869. The tournament began shortly after eleven a.m. on July 5, with the running of hose cart races. Warren Engine Company No. 1 of Carson City brought its fancy carriage for the run, which proved a little tough for some of the firemen. The track went from Young America's engine house to the hall of Knickerbocker Engine Company No. 5, some 1,442 feet down C Street.

Young America Engine Company No. 2 ran first, but due to a disagreement between the timekeepers, was awarded a time of 2:17. While the timekeepers were regulating their watches, Washoe Engine Company No. 4 ran the course and, according to someone's watch, made the run in 1:18. After the timekeepers set their watches, the rest of the competing companies made their bids, with Knickerbocker Engine Company No. 5 posting a time of 1:23; Eagle Engine Company No. 3, 1:26; Monumental Engine Company No. 6, 1:20½; and Liberty Engine Company No. 1, of Gold Hill, 1:29¾. The timekeepers, embarrassed by their earlier disagreement, stated that Young America and Washoe should run again, which Young America did amid strong protests from the other companies, making the distance in 1:18¼. Washoe Engine Company No. 4 was initially willing to run again, but there were so many arguments and such strong language that the situation seemed liable at any minute to turn into a brawl. The company withdrew in disgust and left the silver trumpet prize to Young America Engine Company No. 2, which had sponsored the tournament.

The contest had been a severe test of speed and stamina, and many of the men in the hose carriage race were obliged to quit the ropes before they had run even half the course. Some fell, some were run over, but no one was seriously injured. Eagle Engine Company No. 3 finished the race with only two men out of eight still with the carriage. Only Liberty Engine Company No. 1 finished the race with all eight runners, and even then one would have dropped off and been left behind had he not ridden on the tongue of the carriage to the finish line and a last-place time.

The contest was so disagreeable to those who had run that protests and

Members of the Washoe Engine Company No. 4 hose cart race team pose for a photograph in 1869, prior to running in a hotly contested and protested race along Virginia City's C Street. (Nevada Historical Society)

challenges appeared in the local papers for a full week, and the Board of Engineers of the Virginia Fire Department agreed to sponsor a second running of the race the following week. The results of that race were the same, however, with Young America Engine Company No. 2 reinforcing the grip it held on the prize trumpet.

Following the controversial hose cart races was the grand trial of the hand pumpers, draughting water from the cistern in front of the Sazerac Saloon. Young America No. 2 took the lead, throwing a distance of 200 feet, 4 inches, but was finally beaten by Eagle Engine Company No. 3, which threw a distance of 209 feet, 4 inches. Knickerbocker Engine Company No. 5 won the third-place silver trumpet by hitting just over 109 feet, in the face of an adverse wind. The contests drew a large crowd all along C Street. Balconies were filled with gaily dressed ladies, and sidewalks were crowded with the men of the town. Plenty of whiskey was downed that day, but Police Chief George Downey and his men prevented any serious fighting from taking place.[9]

Few people know the story behind the exquisite silver serving set and trumpet now displayed in the Warren Engine Company No. 1 Museum in Carson City, but it is one of the most interesting stories concerning the firemen of the Comstock, their races, and a bit of romantic coincidence. The story was best told in 1934, in the columns of the *Nevada State Journal*:

> Coincidences! Yes, they happen! Sometimes with a touch of the dramatic and with pleasurable surprise. That is the experience that Miss Clara Crisler is able to tell concerning a silver pitcher and a silver trumpet that after many years have been brought together in her home. Not only a detail of her family history is linked with the episode, but also a chapter of a colorful and former day in Nevada history.
>
> Let us have the story then: We go back 52 years in time. The scene is in Gold Hill. It was a big day in the life of the mining camp when, on a summer day in 1881, the "Butt Enders" and the "Jackets," opposing teams of the local fire companies [Liberty Engine Company No. 1, and Yellow Jacket Engine Company No. 2, respectively] were to run a 450-yard race. The "Butt Enders" won. The prize was a large and handsome pitcher. One would easily misjudge the stuff the volunteer companies of Gold Hill were made of if he thought that the "Jackets" would be satisfied with one attempt. Another race was pulled off. This time the prize was a large silver speaking trumpet with a nice blue cord, tassel and everything. Too bad! The race was to the swift and the honors and

the trumpet became the possession of the "Butt Enders" of Liberty Hose Co. No. 1. An uncle of Miss Crisler, Coonie Pohl, took part in the winning team and an aunt, Florence Pohl, sang the song of victory that had been composed for the big event.

If this were being staged, of course, the curtain would go up at this place and we would call it Act One.

Let the curtain go up now on another and very different scene. A wedding in a mining camp! Miss Minnie Pohl was the bride. That may not mean very much to you, reader, but it meant a great deal to Liberty Hose Company No. 1 for Miss Pohl was a great favorite with them. Often she had entertained them with her ability in song and in a day when the mining camps attracted high talent, she need not take any second place.

What is this coming down the street to Minnie's house? The fire company on parade! The red shirted group with a band playing at the head of the procession. Minnie, you certainly have a surprise coming. Come out here. Then in a nice speech the fire company presents to her the silver pitcher set. There it is nicely engraved. Read it for yourself: "Presented to Miss Minnie Pohl on her wedding day, January 8, 1882, by Liberty Hose Company No. 1 of Gold Hill, Nevada." The silver pitcher thus became a family treasure and in time was inherited by the daughter, Miss Clara Crisler.

But the silver trumpet? With the changing fortunes that befell Gold Hill, the "Butt Enders" disbanded and the trumpet passed from one hand to another. At last it came into the hands of William Greiner of Gold Hill, a one-time member of the "Butt Enders." The thought came to him—and that was a happy idea—to make Miss Crisler a present of the silver trumpet. There was no present that she received at Christmas, 1933, that approached in interest and deep satisfaction the gift of the silver trumpet. Miss Crisler is very sure that there is a Santa Claus.

Thus what circumstance had put asunder, the course of time joined together. After the absence of 52 years, the silver pitcher and the silver trumpet now reside under the same roof in the Capital City of Nevada. They will have an appropriate and congenial company in that home where are gathered so many momentoes, documents and books relating to early Nevada history.[10]

Many years later, the descendants of Minnie Pohl placed the silver pitcher and the silver trumpet together in the museum of Warren Engine Company No. 1 in Carson City. There they rest alongside many photos and other mementos from the Gold Hill Fire Department and the pioneer Crisler and Pohl families.

Minnie Pohl, honorary member of Liberty Hose Company No. 1, was given a large silver set by the firemen and was serenaded by them on the eve of her wedding. Her family was involved in Gold Hill's Liberty Engine Company No. 1, and in Carson City's Warren Engine Company No. 1. (Pohl Collection, Warren Engine Company No. 1)

The silver trumpet was first prize in a hose cart race between members of Liberty Engine Company No. 1 and Yellow Jacket Engine Company No. 2 in the 1880s. The silver pitcher was presented to Minnie Pohl by adoring firemen. Both now repose in the Warren Engine Company No. 1 museum in Carson City. (Author's photo)

There was always something of a love affair between the ladies and the firemen of Virginia City. The ladies were the special patrons of the various fire companies and the firemen were their heroes. The favored ladies came in all sizes and shapes, all ages and social backgrounds. She might be the wife of a mine foreman, or a shady lady from the line down on D Street. She could be the young daughter of one of the townspeople who rode on the hose carriage on the Fourth of July, or a pretty girl like Minnie Pohl, courted by a number of firemen. Whatever her background, the favors she conferred upon the firemen of the Comstock were returned with like ceremony and often a prized memento.

Gifts from the women might be a set of handmade markers or flags. In

1871, Addie Greer presented Eagle Engine Company No. 3 with two pairs of markers made of heavy satin. One set was trimmed in heavy gold bullion and the other in silver bullion. In return she was serenaded by company members and a small band they hired for the occasion.[11] Katy Wilson, a young actress appearing at Piper's Opera House, donated her talents for a performance to benefit Eagle Engine Company No. 3 in 1870. In return she was made an honorary member of the company, and received in addition a gold watch and chain, described as being "appropriately engraved" for its new owner, "a painstaking, deserving young actress."[12]

Lizzie Williams, who was often present at fires working the brakes of a hand engine, stretching hose, or recruiting help, presented a set of silk markers to Nevada Hook and Ladder Company No. 1 in 1868. The markers cost ninety-five dollars for materials, and were made by Lizzie to be presented to the company for use in parades. For her special gift and her extraordinary efforts at fires, she was made an honorary member of the hook and ladder company, and her markers were praised as "the most elegant, valuable and appropriate ones we have yet seen, very creditable as a specimen of her handiwork, and much to be prized as a present to her company."[13]

The wife of John Marks, chief of the Gold Hill Fire Department, was made an honorary member of Liberty Engine Company No. 1 for her help in decorating the hall on various occasions, as well as in decorating the hose carriage for parades. Divide Hose Company No. 2 carried on its roll the names of Susie and Dolly McCone, Mamie Mann, and Etta Simms. They were made honorary members for their efforts in organizing that company's first dress uniform ball in 1890.[14]

An honorary membership was a prized award, but there were other rewards for service to the fire companies of the Comstock. For her gift of a "magnificent pair of markers," made of rich pink satin and fringed with gold, Flora Bray was made an honorary member of Eagle Engine Company No. 3. She was also presented with a gold watch and chain, which bore the inscription, "To Miss Flora Bray, from Eagle Engine Co. No. 3, Virginia, Nevada, July 4th, 1865."[15] In 1867, Young America Engine Company No. 2 received a large American flag and a fine set of silk markers from Amelia Gibson and Maggie Brewer, the latter of whom had recently been married to a man named Dustan. On that occasion the *Gold Hill News* reported, "Mr. James Phelan received these beautiful presents on behalf of the company . . . presentation speeches [were] made by the ladies, and the reception speech by Mr. Phelan." Two weeks

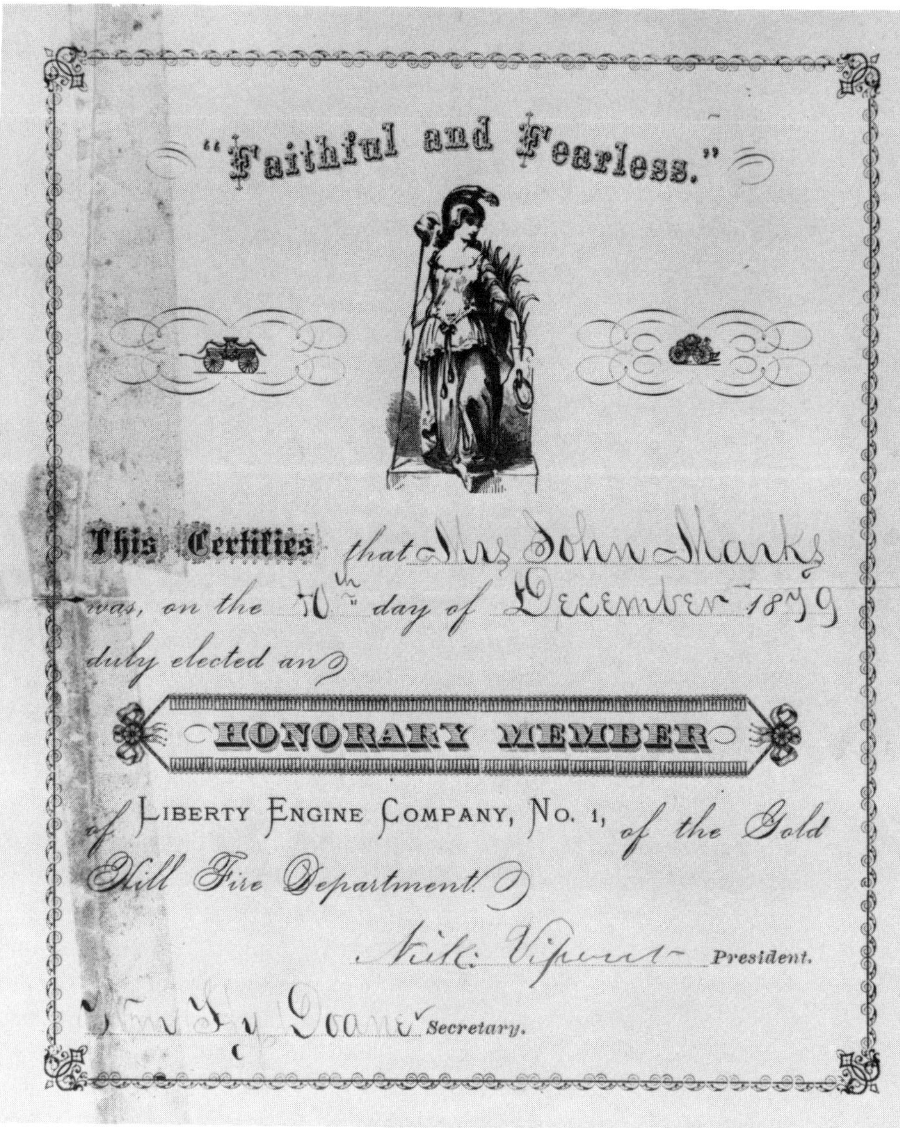

"Faithful and Fearless."

This Certifies that Mrs John Marks was, on the 10th day of December 1879 duly elected an

HONORARY MEMBER

of Liberty Engine Company, No. 1, of the Gold Hill Fire Department.

Nick Vipont President.

Wm F. Doane Secretary.

Meriam Marks, wife of Gold Hill fire chief John Marks, was accorded an honorary membership in Liberty Engine Company No. 1 for her assistance in decorating the company's hose carriage for a parade. Several women were made honorary members of fire companies on the Comstock Lode for their efforts in behalf of firemen. (Bill Marks) later, while the new Mrs. Dustan was dancing at Piper's Opera House, a silver bullion brick was tossed upon the stage for presentation, "to Mrs. A. T. Dustan, by the members of Young America Engine Co. No. 2, Virginia, Nev., December 25th, 1867." The brick became a valued possession for the young flag maker.[16]

Each of the various fire companies in Virginia City and Gold Hill had its favored darlings, but two of the companies had very famous, if very different mascots. Adah Isaacs Menken was one of the most notorious actresses of her day, with a life that newspapers portrayed as rivaling the heroine portrayed in "The Perils of Pauline." "La Menken," as she was billed and known over the globe, had been mistress to noblemen in Europe, and had been married to the prizefighter Benecia Boy Heenan, who taught her to box. She was the type of woman who made tough men weak in her presence. She had been captured by Indians and rescued by Texas Rangers, who taught her to ride horses like a circus acrobat. She was known in Havana as the "Queen of the Plaza," and was hailed in San Francisco as the "Frenzy of Frisco."

It is no wonder, then, that her appearance in Virginia City, in 1864, caused considerable excitement. Giant playbills plastered about town featured a tantalizing picture of her, and promised a finale with Menken lashed to the back of a black stallion, dressed in a scandalous costume of flesh-colored tights and bloomers, topped by a filmy blouse. For her contribution of a special benefit performance for Young America Engine Company No. 2, "La Menken" was made an honorary member of the company and was presented with a Morocco red, silver-mounted fireman's belt, with a large, fancy buckle made of Comstock silver.[17]

Only a few blocks from the opera house where "La Menken" was performing was the small frame house of another famous honorary member of a fire company. This woman was one of the town's soiled doves, one of the best-known prostitutes in the area. Julia Bulette had come to the Comstock in the early 1860s, and became one of the town's most sought-after ladies of the evening. She often catered to the rough element of the town, which probably led to her long affiliation with Thomas Peasley, a member of Virginia Engine Company No. 1, and a man with a reputation for being tough. Her affair with Peasley, and her legendary charitable acts to benefit the fire department, probably led to her being named an honorary member of Virginia Engine Company No. 1. Even when Peasley disassociated himself with Virginia Engine Company and joined Eagle Engine Company No. 3, Bulette retained the red shirt, silver "1" pin, and fancy fire belt and helmet which the members of Virginia Engine Company had presented to her. She was so proud of her fire department affiliation that she ventured uptown from her D Street residence one day to have her photograph taken for posterity.[18]

Comstock society was familiar with the lady from the line and was aware of her special affiliation with the fire department. She was both sought after and ostracized by different segments of the "upstanding" community. But it was her jewelry that caused her undoing at age thirty-five on the evening of January 20, 1867, when John Millain killed her. Millain was arrested for the murder after some of her belongings were found in his possession. He was found cowering under a porch and taken to the city jail. Millain could barely speak English, and his testimony was confused, so his motives and the circumstances of her death were never really made clear.

When Bulette was buried, delegations from the various companies of the fire department turned out in full dress uniform in her honor. They escorted her casket through the town, although not all the way to the cemetery on the small knoll behind the county hospital. However, a line of carriages containing her colleagues in the shady business of D Street, saloon keepers, and perhaps even some previous customers and others who knew her and were not ashamed, did accompany her remains to the lonely City Cemetery plot.[19] A fine fence was erected and the grave marked. In later years she became the heroine of E Clampus Vitus, an organization dedicated to the mining history of the Comstock and to Julia Bulette's memory. Each year a small enclosure is identified as her resting place to thousands of tourists visiting the area, but they are unaware that the grave was moved in the 1950s. Many mistakenly think that the grave of Julia Bulette has been lost, but a handful of Nevadans, sworn to secrecy, know the exact location of her grave in the old City Cemetery.*

Millain's trial was a quick one, with the jury returning a guilty verdict. "When the verdict was announced," Alf Doten wrote in his journals, "the

---

*In 1980, the author was taken to the old City Cemetery by one of those who had moved the fence and marker from Julia's grave so that it might be seen from town, and also so that her remains would not be disturbed. Only a handful of Nevadans know the exact location of the plot, which is near the place pointed out incorrectly as Julia Bulette's final resting place. The City Cemetery has been vandalized, and years of neglect have left it hardly recognizable as a cemetery, with only scattered pieces of wood and a very few monument bases to indicate that it is the last resting place of some of the Comstock's pioneers.

bell of No 1 pealed out in notes of joy—No 2 responded, and some of the steam whistles sounded, thinking it was alarm of fire—some of the engines rolled—Jule Bulette was an honorary member of Va Engine Co No 1 at time of her death, hence the bell ringing."[20]

V. Home Sweet Home

The firemen of the Comstock exhibited a great deal of pride in their "enjines," but besides the pumpers, they were most proud of their engine houses. The companies rivaled each other for the biggest and best-equipped hall in the fire department.

The early engine houses were makeshift affairs. Everything from stables to old saloons were pressed into service to house the engines and hose carriages of the fire department. Shortly after Eagle Engine Company No. 3 was formed in 1863, it moved into the Milton House on B Street, across from where the county courthouse now stands. The Milton House had been leased by the city to serve a dual purpose. The second story was fixed up as a meeting room for the town board of trustees, a city clerk's office, and police courtroom; Eagle Engine Company occupied the lower floor of the building through an agreement with the city fathers.[1]

Eventually the fire companies acquired their own engine houses, some companies moving two or three times before settling into a permanent location. When Virginia Engine Company No. 1 vacated its house adjacent to Piper's Opera House in 1864, the house was acquired by Knickerbocker Engine Company No. 5, who moved it to C Street in the city's First Ward.[2] By the time of the Great Fire of 1875, Virginia Engine Company No. 1, Nevada Hook and Ladder Company No. 1, Eagle Engine Company No. 3, and Washoe Engine Company No. 4 were all located on B Street, while Young America Engine Company No. 2 and Knickerbocker Engine Company No. 5 were housed on C Street, with Monumental Engine Company No. 6 on south C Street, on the Divide between Gold Hill and Virginia City.

The engine house of Liberty Engine Company No. 1, in Gold Hill, was erected in 1869. The bell atop the hall was given to the firemen by Eilley Orrum Bowers after her mill burned down in 1869. A cistern was located under the ramp in front of the hall, and to the left was the hose-cleaning and storage room. The building stood until heavy snows collapsed it in 1952. (Author's collection)

Although the firemen of Virginia City and Gold Hill needed little excuse for a celebration, the completion of a new engine house was a special time, prompting speeches and frivolity. When the Liberty Engine Company No. 1 hall was completed in Gold Hill in 1869, the *Gold Hill News* pronounced the "housing" of the engine in the new hall "one of the gayest and most festive arrangements on record. Of eatables and drinkables there was no lack, and as for true jovial conviviality, there was no end to it. The new engine house was crowded to its utmost capacity, and a great crowd filled the street in front." There were some speeches, readings from the Bible, and several hymns. The *Gold Hill News* reported:

> Stacks of bottles were emptied, and although it is said nobody was drunk, everybody was just as happy as a clam at high water. Nye was the slyest old rat in the crowd, and when he found the beverages coming too fast upon him he slipped off to bed, but the boys got after him, and under the pretence of requiring him to pacify Lee, whom they represented to be on a rampage, ready to sink, burn and destroy everything and everybody, they got Nye back to the house, where he had to do some responsible speech making and drinking before he was let off again.*[3]

The engine house of Liberty Engine Company No. 1 had some very unique features. In addition to the hall and meeting room downstairs, the garret contained several rooms suitable for use as living quarters for members of the company. This firehouse was probably the first on the Comstock to feature a special adjacent building for the purpose of cleaning, oiling, and storing the leather hose of the company after a fire. Elsewhere this process took place on the boardwalks in front of the various engine houses, much to the consternation of passing ladies, who disliked oily spots on clean clothes. Another feature unique to the Liberty Engine Company hall was the cistern located in front of the large doors, just under the engine ramp. This allowed firemen to pump the hand engine from inside the hall in the event of a nearby fire, something that was proudly done by the "Butt Enders" on several occasions.[4]

Each of the engine houses, whether for the hook and ladder company, a hose company, or an engine company, was uniquely decorated to suit the tastes of the company members. Leather helmets and caps were displayed on fancy wooden racks on the walls, with brass parade

---

*Peter Nye designed the Liberty Engine Company No. 1 firehouse and was overseer of its construction. W. H. H. Lee was chief engineer of the Gold Hill Fire Department and a member of Liberty Engine Company No. 1.

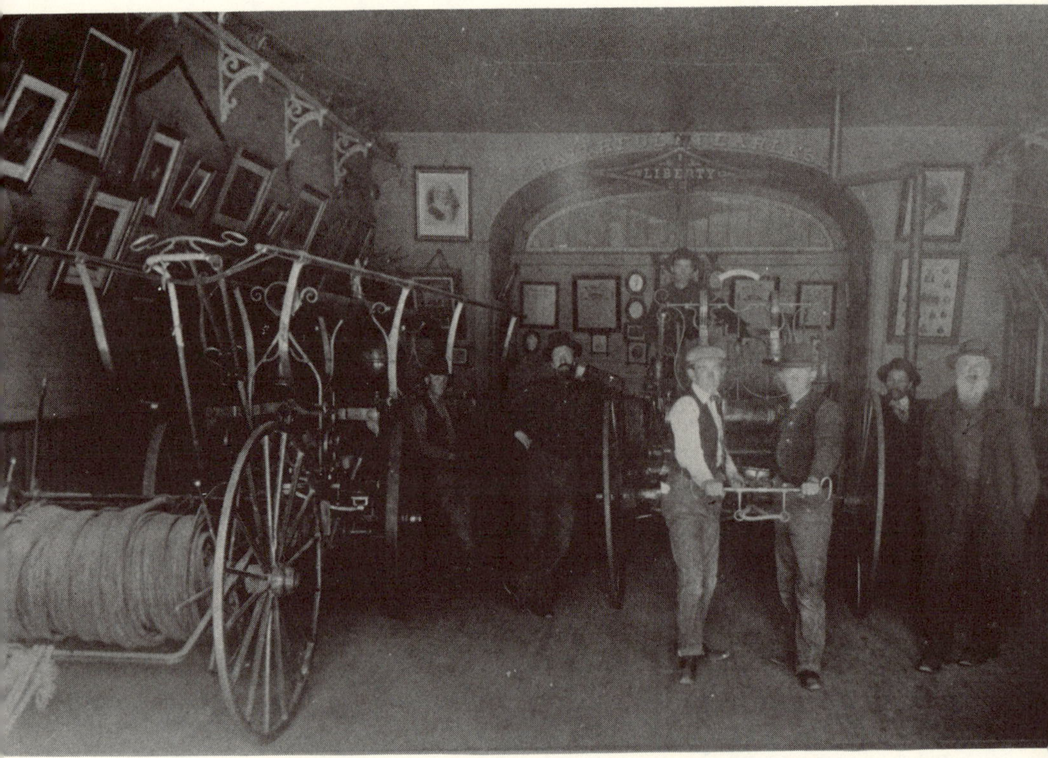

This rare view of the interior of the Liberty Engine Company firehouse in Gold Hill shows a hall decorated with memorabilia as well as the hand pumper, four-wheel hose carriage, and two-wheel jumper belonging to the company. Picture probably taken in the late 1880s. (Comstock Firemen's Museum)

torches, axes, belts, and other items under them. In the corners and elsewhere around the hall were silk banners, American flags, and presentation markers bearing the name and/or number of the company.

Ornaments and decorations for the engine houses were presented to the various companies by persons in the community. Copies of famous paintings, original works by local artists, portraits of famous people, and other items adorned the halls of the various volunteer fire companies of Gold Hill and Virginia City. Liberty Engine Company No. 1, in Gold Hill, was presented with a complete set of color lithographs of the 1854 Currier and Ives series known as the "Life of a Fireman." The series depicted firemen in heroic scenes of daring rescues, equipment at fires, and companies rushing to fires.[5] Eagle Engine Company No. 3 displayed

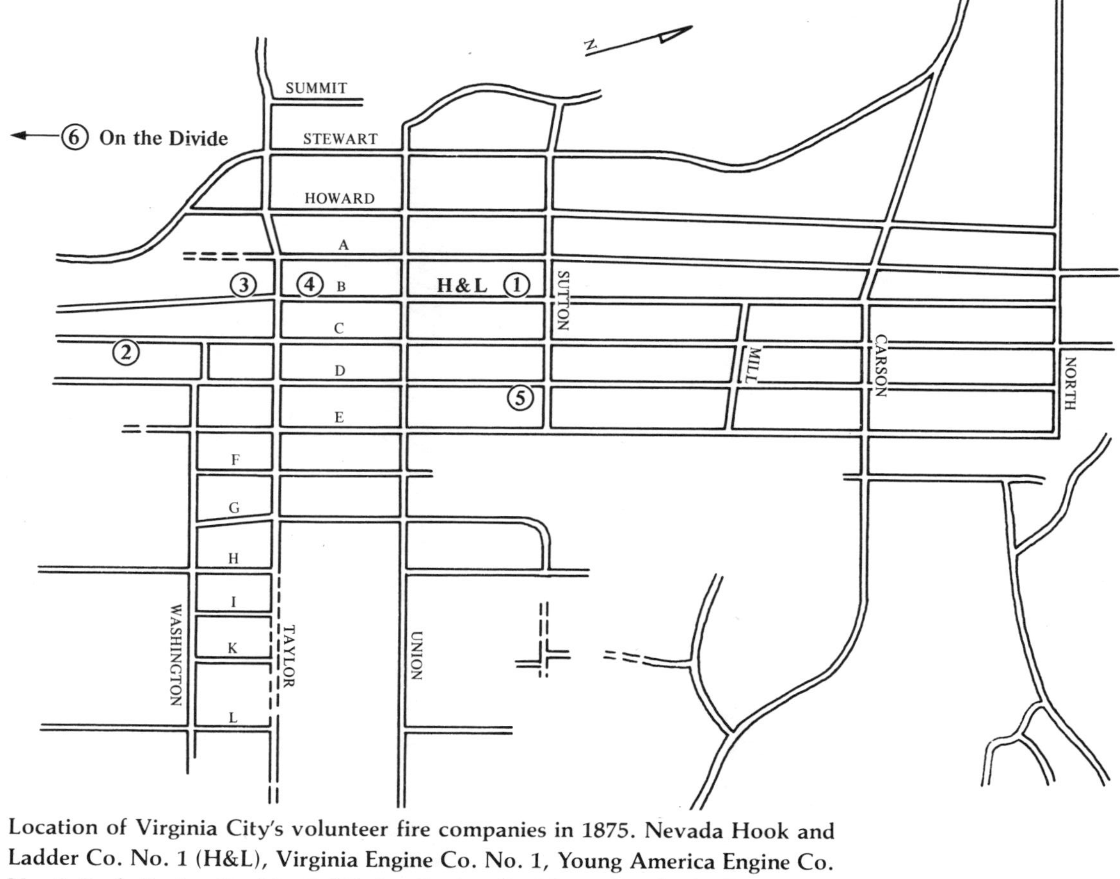

Location of Virginia City's volunteer fire companies in 1875. Nevada Hook and Ladder Co. No. 1 (H&L), Virginia Engine Co. No. 1, Young America Engine Co. No. 2, Eagle Engine Co. No. 3, Washoe Engine Co. No. 4, Knickerbocker Engine Co. No. 5, Monumental Engine Co. No. 6.

The two-story brick engine house of Young America Engine Company No. 2 was one of the finest in Virginia City. Although it sported a fancy bell tower, the large bell ordered by the company had to be moved to the center of the roof because when it was rung in the belfry, chunks of plaster would fall inside the engine house. (Bob Dockery)

a special wreath of flowers in the form of a figure three. It was described as "about three feet in length, and the white roses, green leaves and silver wheat heads combine so harmoniously, that the whole effect is an exquisite blending of colors that is delightful."[6] Such special presentations were surrounded by all the trappings of the fire department, ornamental and symbolic as well as functional.

Young America Engine Company No. 2 had one of the finest engine houses in Virginia City. Its two-story brick hall, on C Street opposite the Presbyterian church, was illuminated by fancy gas fixtures. It had fine oak chairs for the members of the company, and a giant oak desk for the use of the secretary and foreman of the company. Upstairs was a furnished reading room with several hundred volumes of books, including the works of Shakespeare, Milton, Ben Jonson, Byron, Scott, Burns, and a score of others, as well as the leading newspapers and periodicals of the day.[7] Inside the hall itself were the silver trumpets of the officers, or those won during hose cart races and pumping contests, brass bells, ornamental eagles, and fancy lamps. Also hanging in a special place of honor in No. 2's house was a hand-drawn scroll incorporating photos of the membership, made in 1869 by W. M. Metcalf of the firm of Thompson and Metcalf, which specialized in penmanship and bookkeeping. The company's roll was praised in the press as "a very fine and artistically executed piece of penmanship," featuring a drawing of the engine house with the engine in front, and ornamentally designed spaces into which a photograph of each member was to fit. "The whole forms a conspicuous ornament to the Hall, and is a credit to the artist," the newspapers wrote.[8]

The companies also took great pains in decorating the exteriors of their engine houses. After Eagle Engine Company No. 3 moved to its new house on south B Street, adjacent to the offices of the Virginia & Gold Hill Water Company, the members began an extensive fix-up campaign, including a gleaming coat of whitewash. The firemen did such a good job in cleaning and painting their new engine house that many people mistook it for a church, which greatly annoyed the company. As a result, the company had its name and number painted on the transom over the door.[9] The house of Lincoln Hose Company No. 3, in lower Gold Hill near the Rhode Island Mill, was, in fact, a house of worship. In 1879, Lincoln Hose purchased the Methodist church building in Gold Hill from the Roman Catholic Church, for use as a firehouse. J. F. Gladding, who became chief of the Gold Hill Fire Department in 1885, won the contract for moving the building and fixing it up as a suitable home for the company's carriage and equipment.[10]

Each of the companies had a steward to maintain the hall, ring the alarm bell, sound the call for meetings, keep the engine and hose carriage cleaned and polished, and perform other tasks set by the company officers. The stewards were generally members of the company, elected for a year's term, and entitled to live in the firehouse. They were paid—sometimes by the company, sometimes by the city or county, depending upon individual agreements. Humor often surfaced during the election of the company steward, as during the election of Young America Engine Company No. 2, in 1867, which was reported by the *Gold Hill News*:

> Young America Engine Co. No. 2, Virginia, is a big institution. They are a very efficient, harmonious and frolicsome set of fellows—the members of No. 2. "Sam" [Wyckenheim] is the Steward. At the last election Sam had a very hard race for re-election—some half a dozen competitors. Sam restled. Geo. Smith of Red Mountain squeezed himself in as one of the candidates. Indeed, Sam had doubts about his success. When the election evening came, a resolution was proposed to the effect that the successful candidate should pay the "lager" for the crowd. Sam pushed hard for passage of this resolution; and when it was passed all of his rivals withdrew from the field. Sam paid the lager with a hurrah. Since then it has occurred to him that something was "put up."[11]

The Board of Aldermen often paid the salaries of the stewards and also provided supplies of wood and coal for the companies during the harsh winter months. The stewards were charged with keeping stoves going to prevent the engines from freezing, and thus having them ready to respond to the alarm of fire during any season.

Although the engine houses were generally not the scene of annual balls and other major celebrations, they could become the scene of a spontaneous party, such as that held by Knickerbocker Engine Company No. 5, in 1871, after firemen returned from an especially hard-fought fire. It was a "grand jollification," with food, songs, and plenty to drink. Although they might be rivals, it was not uncommon for one fire company to invite the members of another into their hall for a drink of "lager" after a particularly tough blaze, and it was not unusual for a table of "sumptious eatables and drinkables" to be put out in the engine houses for members and others who worked at the fire.[12]

The only known full view of the engine house of Yellow Jacket Engine Company No. 2, Gold Hill. The company's Button & Blake hose carriage is in the doorway; it is now on display in the Comstock Firemen's Museum in Virginia City. The men are wearing leather helmets and rubber turnout coats. (Warren Engine Company No. 1)

The fire bells atop the various engine houses were almost as colorful as the men who rang the alarm of fire. Although the steam whistles of the mills and mine hoisting works often signalled the alarm of fire with shrill blasts that pierced the night, the clarion tones of the bells on the engine houses were the universal rallying signal for the men who wore the red shirts and leather helmets. Members of the respective companies of the fire department were summoned to their engine houses to conduct business by a light tap of the bell. The bells also signalled gatherings to drill with the engines, or meet with another fire company in a contest of the squirting prowess of the "masheens."

The fire bells were also used to toll in mourning of the passing of members of the fire department. When John Dunn died, the bell of Young America Engine Company No. 2 began to toll at about seven a.m., shortly after the death was announced.[13] The bells tolled in honor of prominent citizens of the community, even though they might not be affiliated with any particular company or with the fire department itself. The bells also signalled important news to the firemen, such as the conviction of John Millain for the murder of Julia Bulette.[14]

The bell of Young America Engine Company No. 2 had a varied and somewhat dark history. The engine company, always seeking to be the best and have the finest equipment, purchased a giant bell in San Francisco. It had served Monumental Engine Company No. 6 of that city, as well as having sounded the call to arms to the vigilantes atop "Fort Vigilance." The huge bell proved troublesome for the men of Young America, however, after it was placed in the cupola atop the engine house on C Street. When the heavy alarm bell was rung the first time, it shook and jarred the building, rattling down the plastering inside, and cracking windows. As a result, the bell was moved toward the back of the building where, although less conspicuous, it could do less harm.

The bell's history of vigilance participation did not end when it was brought to Virginia City. When it was put up for sale in 1877, the *Territorial Enterprise* reminded its readers that, "Here it was used by '601,' and on a certain occasion when the steward attempted to ring it, it was found to have been muffled." The bell was sold in 1878, after the disbanding of Young America as a fire company. When it was taken from its lofty perch on the engine house, the *Territorial Enterprise* decried the fact that the workmen lowered the historic bell "without allowing it to sound a single note of regret."[15]

Some of the fire bells of the Comstock travelled quite a bit. The clarion tones of the bell atop the Liberty Engine Company No. 1 firehouse in Gold Hill originally sounded shift changes at the Bowers Mill in Gold Hill. After the mill burned in 1869, the "Washoe Seeress," Eilley Orrum, presented the bell to the Butt Enders for their new engine house. The bell sounded the alarm of fire in Gold Hill until 1938, when the company finally disbanded. It was saved from the wreckage of the firehouse after it collapsed under the weight of heavy snows in 1952, and today stands on the original site in its old cupola, a monument to the pioneer firemen of Gold Hill.[16]

When it arrived in 1869, the bell of Monumental Engine Company No. 6 was welcomed by the *Gold Hill News* with the comment that it "has a clear, sharp tone sufficient to wake the dead, in case a 'wake' was in order." Previously, the citizens of the Divide, between Virginia City and Gold Hill, could not always know of a conflagration in either of the towns without a general alarm of all bells and steam whistles in the towns. With the addition of the new bell, the newspaper noted, the degree of "get-up-and-git-ativeness" of the Monumental members would "doubtless be much increased should their new bell call on them in the night."[17]

On New Year's Eve of 1876, this bell was rung for so long in greeting the centennial year that it cracked. The 250-pound bell was hauled down and a new one ordered cast. John Kennis, a member of the fire company and an expert bell caster, bossed the job of casting Monumental's new bell, which was to be something special, with forty dollars' worth of silver bullion added to the casting metals courtesy of mining magnate John Mackay. When the new bell was raised and tested, the *Gold Hill News* proclaimed, "The tone of the bell is as good, perhaps finer than any other in the city."[18] When the Monumentals disbanded in 1879, the bell was purchased by Lincoln Hose Company No. 3, and placed atop the former church serving as their hose house to sound the alarm in that section of the Comstock.[19] Many years later, after Lincoln Hose Company No. 3 ceased to exist, the bell was given to the Nevada Historical Society in Reno, where it was displayed until 1981, when it was loaned to the Comstock Firemen's Museum and returned to Virginia City.

The fire bells often had a humorous application as well as a functional one. A red-faced Gold Hill Fire Chief Nick Sexton found such an occasion on his wedding night, when the bell of Lincoln Hose Company began the alarm of fire, and was soon joined by the other companies of Gold Hill. When the fire chief left his wedding bed to turn out to the fire, he found that a few of the "boys" had gotten together, and, bolstered by courage from "a couple of kegs of beer," thought to start their chief's marriage off right.[20]

Perhaps the most beloved fire bell of Virginia City was that of Virginia Engine Company No. 1, a bell which sounded the alarm of fire in Virginia City from 1861. It was the first fire bell in Virginia City, and thus the state's first fire alarm bell. It was cast in England and brought to the Comstock during the earliest days of the fledgling fire department. The large bell was moved several times, as Virginia Engine Company No. 1 sought bigger and better quarters over the years, and survived at least four major fires, including the Great Fire of 1875. Following that conflagration, which destroyed the engine house of Virginia Engine Company No. 1, the bell was found to be as good as new, despite exposure to fires and winter storms. It was decided to suspend the bell over the new engine house of No. 1, "and as in the days of old [it] will lend its voice to send the alarm in case of fire."[21]

When Virginia Engine Company No. 1 disbanded with the rest of the old volunteer fire department, it sold all its property, including the bell, to the city, for $2,750 in gold coin. When the new Corporation Fire House was completed and the new fire equipment housed, the old bell was moved there rather than having the city go to the expense of purchasing a new one.[22] The bell remained on the Corporation Fire House until 1934, when the fire department moved into an old saloon building on C Street (now housing the Comstock Firemen's Museum) to facilitate modern motorized equipment. When the present firehouse was completed in 1962, the state's oldest fire alarm bell, once sounding the alarm for the hand engine of Virginia Engine Company No. 1, moved again. Today it sounds the curfew alarm nightly, with ghostly but clarion tones as sharp as the day it was first hung on the Comstock to rally the men in red shirts and leather helmets.

VI. Scandal

Scandal could and often did rock the volunteer fire departments of the Comstock, sending the wags of the town up and down the boardwalks offering their own interpretation of the current gossip. Most of the fire department scandals centered on individual company funds, on funds controlled by the officers of the fire department, or on the charitable fund, which was controlled by the city recorder. Sometimes, however, there was a twist of humor in scandals involving the firemen of the Comstock.

In 1865, City Recorder William Davenport was charged with embezzling funds paid to the Charitable Fund of the Virginia Fire Department, and few firemen could find anything humorous in the situation. The Virginia Board of Fire Delegates referred to Davenport's sneak theft as "acts unworthy a public officer, an honest man or a gentleman." The board further passed resolutions, which were printed in the newspapers, promising, "We will use all honorable means to defeat the said Wm. H. Davenport, in the event of his ever presuming to again become a candidate for any office." The board concluded that Davenport was "unworthy the public confidence, and we leave him with contempt to enjoy his ill-gotten gains." Davenport, who had served as judge in the city courts, vanished before the anger of the volunteer firemen could be felt. If he ever had thoughts of returning to the Comstock, they were certainly quelled by the vehemence of the resolutions against him.[1]

A less serious scandal involved Chief Peter Larkin, at about the same time as the discovery of the funds missing from the charitable fund. Chief Larkin, who was in charge of insuring that the various fire companies were outfitted with necessary tools and equipment to handle any emergency, apparently felt it his duty to equip each company's foreman with a brass or nickel trumpet with which to bark orders to the men during fires. Trumpets for each company were ordered from Gillig, Mott & Company, one of Virginia City's most prominent hardware stores.

The trumpets had already been delivered when the bill for them arrived, and the Board of Delegates charged that Chief Larkin had overstepped his authority in ordering them. The fire chief was furious at this affront to his powers and his actions, but under pressure from several critics on the board, he offered to pay for the questioned trumpets personally. The Board of Delegates took the matter under advisement, apparently letting Larkin stew a little over the possibility of having to buy several very expensive trumpets. In the end, however, the delegates issued instructions to the various fire companies to either return the trumpets or pay for them out of company funds, refusing responsibility for buying the trumpets with fire department money. Larkin was off the hook, but his face was red for some time.[2]

Although not frequent, financial scandal continued to trouble the fire departments of the Comstock late into the century. In 1886, the treasurer of the Gold Hill Fire Department, J. A. Winterbauer, was charged with financial misconduct for failure to pay warrants "drawn on him by the department." A special meeting of the Gold Hill Board of Fire Delegates was convened in the hall of Liberty Engine Company No. 1, of which Winterbauer was a member. President John Wadge called the meeting to order, noting the absence of "J. A. Winterbauer, the suspect," as well as several other members of the board. Otto Lutgens, a delegate from Lincoln Hose Company No. 3, in lower Gold Hill, preferred the charges against Winterbauer on behalf of the fire department. A copy was ordered sent to the suspect treasurer, informing him of the charges and setting a time for a formal hearing on the matter.

The trial was held on a hot August evening in the hall of Liberty Engine Company No. 1. After the secretary of the fire department produced the unpaid warrants, a tense J. A. Winterbauer stood to address the gathering:

> In palliation Treasurer Winterbauer stated that the money to pay the said warrants had been laid aside and subsequently he had forgotten where it was. Having no more to say, Treasurer Winterbauer was requested to retire while the case should be decided by the board. But, there was no deliberation in the matter. Winterbauer was found guilty and summarily dismissed from office. A

new treasurer, Charles Tobener, was immediately sworn in with the stipulation he file a $200 bond. After all was said and done, Tobener reported to the board that, "After examining the account of the late treasurer, I find that he is indebted to the Fire Department in the sum of $15.70.[3]

Tobener served as treasurer for several years, watching the fortunes of mining on the Comstock take a steady downhill path, and the ranks of firemen shrink a little more each year. In 1889, Tobener was again elected treasurer, but was never sworn in. After three years as a trusted treasurer with a bond, he argued before the Board of Delegates that the two-hundred-dollar bond should be waived. After all, he told the board, times were hard and money tight. The members of the board adamantly refused to put any temptation in the treasurer's path, and Tobener stepped aside as C. J. Waider was elected in his place. The new treasurer paid the required bond quietly.[4]

Hard times continued to plague the Comstock, contributing to even more financial scandal in the Gold Hill Fire Department. In 1910, there was considerable discussion about the possibility of selling the old hand engine of Liberty Engine Company No. 1. A meeting was called, but only a handful of members appeared to decide the fate of the old pumper. One member felt it was important to have the rest of the fire company on hand to decide the issue, but the minutes show that President Tim Stack was insistent. "If any of the members had any fault to find it was their own fault for not attending the meetings." Stack pressed for sale of the old hand engine and won. It was finally sold for $450, considerably less than the $1,800 offered for it only three years prior by Bishop Creek, California. The $450 price tag included a sizeable commission for the man who found a buyer for the old engine of the company.[5]

Still one more scandal surfaced in Liberty Engine Company No. 1 before it faded from history. In 1927, times were extremely hard, with mining at a virtual standstill and America on the verge of the Great Depression. The minutes of the company meeting on February 8 of that year show that "Mr. Pollard stated that some of the members were out of work & were in need and suggested that $15.00 be allowed to those needing same." The motion was passed and five members took advantage of the draw from the treasury. At the March meeting, however, the draw caused a controversy, as other members charged that "certain members got up a meeting and had checks written for themselves." After considerable debate, which was heated at times, the secretary was notified to fill out a new roll of the company, with the money remaining in the treasury to be divided among the company members still residing in Storey County. It was to be the last meeting of the company noted in the minutes book until August 12, 1930, when regular business meetings were again recorded.[6]

Financial scandal was no laughing matter. It could produce serious and damaging results for anyone involved, but other forms of scandal sometimes had a humorous side. There was a saying on the Comstock that was applied to husbands straying from the righteous path of a good, faithful mate; and it played a critical role in one Virginia City incident. The *Carson Daily Appeal* tried to explain the origins of the saying, "Oh, he's on the roof!" with the following tale of an incident involving Washoe Engine Company No. 4 and one of its members:

> Years and years ago a leading member of a fire company became enamored of a married lady, and she reciprocated the jakie's attentions until his frequency at her house became a public scandal. The whole town knew of the liason except the trusting husband. While the husband, who was also a member of the company, was in the mine toiling for his bread, the rival of his wife's affections was whiling away the hours with the faithless one on B Street.
>
> One day the adjoining house caught fire. It was but a step from one roof to the other. The bell sounded the alarm, and in a few seconds the two houses, both threatened, were surrounded with a mob of people.
>
> The guilty fire jakey had lingered too long, and as he peeped through the closed curtains of the darkened room he saw that the three o'clock shift was off, and the husband was rushing up the street to save his threatened property.
>
> But the man was equal to the emergency. Grabbing a pail filled with water he carried it through the side window out on the roof below in his shirt sleeves and shouted:
>
> "This way, boys! Look lively now," and then clearing the intervening space at a bound he dashed the water on the blazing roof and shouted at the top of his lungs:
>
> "First water for Fours!" Every fireman below him had figured out the real situation in an instant (all except the husband, of course), and a grand shout went up in recognition of the feat of firemanship. Then he threw off his vest and began to yell for more hose, and for the next 10 minutes, worked like a quarterhorse, the crowd cheering his efforts lustily.

After the fire the husband thanked him feelingly for saving his buildings, and with tears in his eyes said:

"Al, if you hadn't been there on time I wouldn't have a roof over my head tonight."

"You bet I was there Bill," was the reply. "When I can help a neighbor, I'm there."

The next day's edition of the *Territorial Enterprise* carried an account of the fire which included mention of the incident. "Al — — — was the first gallant fireman to scale the roof, and but for his daring and tireless work the whole block might have gone. The families in the neighborhood all breathed prayers of thankfulness last night that Al was there on time." It took a little time for the joke to make the rounds of the town, but before long the phrase, "He's on the roof," began to circulate as the description of anyone wayward in his amorous pursuits.[7]

Chief John Marks: Acts of Sedition

JOHN MARKS was a good fireman, well-liked not only by the members of his own company, but also those of the rival companies in the Gold Hill Fire Department. When he was married, large delegations from the various fire companies turned out, and the newlyweds were serenaded by the Gold Hill Brass Band.

Marks was known as a direct man who seemed to have a knack for gettings things done, although he was a little gruff at times and did not pay much attention to protocol or diplomacy. Those were characteristics that won him membership in Liberty Engine Company No. 1. They were also characteristics that resulted in serious trouble for him in 1883, during his first term as chief engineer of the Gold Hill Fire Department.

It was no surprise that "Johnny" Marks was elected chief. He had shown his competence and devotion to the fire department many times, and had held many important offices. In 1871, he served as delegate from Liberty Engine Company No. 1 to the Gold Hill Board of Fire Delegates. In 1873, he had the distinction of being the first fireman to be elected, unopposed, to the newly created office of second assistant engineer of the fire department. He retained that position the next year, however, by only one vote.[1]

Just prior to his election as first assistant engineer in 1875, Marks was presented with a gold-mounted silver badge, as a token of friendship from Chief Sandy McMartin, a member of the rival Yellow Jacket Engine Company No. 2. Yet despite his popularity with the firemen of Gold Hill and his abilities as a fireman, John Marks had a stormy first term as chief engineer of the Gold Hill Fire Department.[2]

Shortly after his election, Marks devised a plan for the reorganization of the fire department. The plan would have taken some of the independence from the volunteer companies, while making the chief engineer a virtual dictator. The fatal flaw in the plan, however, was the fact that Marks failed to bring his proposal before the Board of Fire Delegates before presenting it to the county commission. On May 18, 1883, only a month after his election, Chief John Marks met a room full of angry firemen. Although upset, the firemen were willing to allow Marks to withdraw his plan so the board of delegates could at least discuss the proposed reorganization, but Marks stubbornly stuck with both his proposal and his methods of presenting it.[3]

By June 1, there was more than a little anger in the hearts of the delegates from rival Lincoln Hose Company No. 3, of lower Gold Hill. They brought formal charges against the chief, resulting in his suspension pending trial before the Board of Fire Delegates. The charges read in part:

> We the undersigned members of the Fire Department do hereby prefer charges against the present chief engineer of the department and do charge the said Chief Engineer John Marks with sedition and ungentlemanly conduct towards the said department and by divers [*sic*] and sundry acts has maligned and belittled and has attempted to break up and destroy the Department to subvert his own personal aims and ambitions.[4]

Twice the delegates met to prefer the charges against Marks, but he refused to attend the meetings, meet with the board, hear the charges, or defend his actions. Had not a delegate from Liberty Engine Company No. 1, J. A. Winterbauer, stepped in, Marks's career as chief engineer might have been a very brief one. Winterbauer proposed a motion to reinstate the chief engineer without hearing the evidence against him.[5]

After some heated discussion, the delegates concluded that the chief engineer had been sufficiently informed of the feelings of the fire department and, thus, put in his place. The motion was passed. Marks dropped his proposed reorganization plan and resumed his duties as chief engineer of the Gold Hill Fire Department. The rest of his term was relatively calm. By election time the following April, the trouble had apparently been forgotten, and Marks was elected to a second term. He was even presented with a special gold-mounted silver badge bearing his name and title.[6]

John Marks was a member of Liberty Engine Company No. 1 for more than twenty years, and although he did not serve as an officer in the Gold Hill Fire Department after 1885, he held the important office of foreman of Liberty Engine Company No. 1 from 1887 to 1893.[7] He is buried in the southwest corner of the Gold Hill Cemetery, where his descendants from Virginia City still care for his grave and that of his wife, Meriam.

VII. *Rivalries and Friendships*

A Nevada fireman of the 1980s remarked that those in the fire service share a camaraderie different and stronger than any other profession, organization, or fraternity. The bonds between firemen span the globe and minimize the barriers created by differences in departments or companies, race, color, or political affiliation. This camaraderie is an ancient tradition, and was certainly present during the days of the red shirts and leather helmets on the Comstock. Enduring friendships were created as one company extended its courtesies to another.

These ties began simply, as in 1865, when the chief of the Virginia Fire Department invited Yellow Jacket Hose Company No. 2 and Liberty Hose Company No. 1, of the Gold Hill Fire Department, to participate in the Virginia City Fourth of July celebration. The "boys" of Liberty Hose unanimously accepted an invitation from Eagle Engine Company No. 3, to be their guests on the national holiday, while "Jackets" were hosted by Virginia Engine Company No. 1. "Both our companies will be the honored guests of Virginia Firemen on the coming day of a national jollification," the *Gold Hill News* proudly announced. Such invitations usually meant that the guest company marched in the parade with their host, and participated in contests with the firemen from the host company. Following the festivities, it was not uncommon for the host company to hold a reception at their engine house and serve lager, wine, cheese, and crackers.[1]

Holidays and special celebrations were always a time to renew friendships with firemen of the other companies of the town, or with firemen from other towns. Visiting firemen were treated like visiting dignitaries. When the cornerstone of Nevada's capitol was set in 1870, the firemen of

Carson City requested that the fire departments of Virginia City and Gold Hill send representatives to attend as their guests. Eagle Engine Company No. 3 of Virginia City and Liberty Engine Company No. 1 of Gold Hill were selected to represent their respective fire departments. With their engines they journeyed on the Virginia & Truckee Railroad to Carson City, where they attended the ceremonies on the capitol grounds and later festivities in the engine houses of their hosts, Warren Engine Company No. 1 and Curry Engine Company No. 2.

"After the cornerstone was laid the fire jakies had some fun," the *Gold Hill News* reported:

> The Curry's squirted, and sent a horizontal stream 205 feet. The Eagles then tried it on, and couldn't fetch that far by five feet or more. At another cistern the Warren's went to work. They threw 210 feet. The Liberty's didn't want any of it, and declined to play. This all recapitulated means that the Curry's showed a longer stream than the Eagles, and that the Warren's showed five feet more than the Curry's. Beat the crack Virginia engine pretty badly.

The spirit of the contest was friendly, however, and shortly after the squirting trial, Liberty Engine Company No. 1 extended a special invitation to the members of Warren Engine Company No. 1 to participate in Gold HIll's Fourth of July celebration. Through the courtesy of the Virginia & Truckee Railroad, a special train was run from Carson City for the Warrens and their engine. More than one hundred members of Sacramento's Howard Engine Company No. 3 also joined the festivities in Gold Hill. It was a homecoming of sorts for some of the members of Howard Engine Company. Many of them were former residents of Gold Hill, and several had served in the fire department, among them W. H. H. Lee, who left Gold Hill to become chief of the Sacramento Fire Department.[2]

Putting "first water" on a fire was very important as a display of the fire companies' efficiency and prowess. It was exciting for the firemen to race down the street, beat their rivals, and get to work in putting out the fire. But two or more companies might claim the honor at a given fire, sparking controversy and stoking the already burning fires of rivalry. When a stovepipe caught fire in Gold Hill in 1865, it was extinguished with a few buckets of water, and a member of Liberty Hose Company No. 1 claimed that the company had "splashed first water." However, the newspaper account angered the members of rival Yellow Jacket Hose

Company No. 2, who, according to the newspaper, couldn't get water into their hose before the fire was extinguished. The Jackets virtually assaulted the reporter, who was apparently a bit red faced at having been bamboozled into reporting the Liberty version of the fire:

> Mr. Wagner, a member of Liberty Hose who yesterday gave us the item about the fire near the Granite Mill, was slightly mistaken when he told us that the fire was out when Yellow Jacket Hose commenced playing on it. The contrary is true, as Mr. Aylsworth informs us. He ought to know, as he is foreman. But for the very timely arrival of the Yellow Jacket Hose there would have been a very serious fire and considerable damage done, as the row of houses where it occurred are mostly frame buildings. In justice to the Yellow Jacket boys, who are ever ready to do their duty, we make the correction.[3]

Another source of continued fighting between the two companies involved creation of the Gold Hill Fire Department. The proposal called for the property of the fire companies to be deeded to the town fathers, to prevent the kind of firemen's strike which had crippled the fire-fighting capabilities of the Virginia Fire Department sometime earlier. Other clauses in the proposed ordinance were considered obnoxious for one reason or another, and the members of Liberty Hose Company No. 1, the town's pioneer fire company, vehemently opposed the ordinance.[4]

On October 2, 1865, members of Liberty Hose Company No. 1 met to discuss the proposal to form a fire department, a proposal many of the firemen considered premature. After lengthy discussion and several arguments, the company voted to seek repeal of the ordinance, in order to allow representatives from Liberty and Yellow Jacket to discuss a draft that would be mutually satisfactory. If no agreement could be reached, they decided, the companies should be allowed to continue to work independently, without an ordinance. N. A. H. Ball, of Liberty Hose, described that company to the town board as "comprised of members who represented the varied interests of the town and had the general welfare of the town at heart." But, he told the town fathers, the company was definitely against the proposal establishing a formal, organized fire department under the stipulations that had been written into the ordinance.

Obviously, a compromise was necessary for the town to have adequate fire protection. The town board had to make some sort of sacrifice to have its version of the fire ordinance passed without deletion of several critical passages, including vestiture of the fire equipment in the town. Finally, new hose was promised for both fire companies if they would relent and endorse the ordinance, which they promptly did.[5]

But the promised new hose was to cause hard feelings between the two companies. When the hose arrived, Yellow Jacket Hose Company No. 2 went to the freight depot and claimed 700 of the 1,000 feet which had been delivered. When they found out, the Liberty members were ready to do battle. After several days of negotiation, punctuated with threats to disband the new fire department, as well as the companies themselves, Jackets agreed to give up 300 feet of the new hose, and the *Gold Hill News* announced that the Liberty firemen had taken delivery of the contested hose, "and have got it fixed up and put on their truck ready for use."[6]

Even though rivalries were sometimes heated, and harsh words were exchanged by members and officers of the various fire companies, there were times when friendships meant more than winning a hard-fought contest over who got "first water." On October 14, 1874, Gold Hill Fire Chief Alex McMartin presented specially ordered gold-mounted silver badges to his first and second assistants, Charles E. Nuttall and John Marks, as a token of his friendship. Both McMartin and Nuttall were members of Yellow Jacket Engine Company No. 2, but Marks was a member of rival Liberty Engine Company No. 1. The presentation took place in the hall of Yellow Jacket Engine Company, following its regularly scheduled meeting. *The Gold Hill News* reported the ceremony the next day:

> Mr. Nuttall received his badge at the hands of Mr. M. P. Wolf who made one of his characteristically excellent speeches. Mr. Nuttall responded, thanking Mr. McMartin very heartily for the beautiful souvenir, and insuring him that it should always be worn next to his heart. Mr. A. J. Graham was called upon to inform Mr. Marks of the good feeling entertained toward him by Chief McMartin, as evidenced by the beautiful badge bearing his name. Graham was equal to the emergency, and made a very neat and appropriate presentation speech, which was responded to very feelingly by Mr. Marks. The presentation exercises over, the whole party adjourned to Dumars' saloon and indulged in a little social fraternization. Numerous toasts were drank to the health of "Sandy," a better chief and larger-hearted man than whom cannot be found.

The death of a member of a fire company, under any circumstances, was an occasion when rivalries were laid aside, as members of the various companies gathered to march together in the funeral procession.

Ben Ballou was assistant foreman of Eagle Engine Company No. 3, and very active in the affairs of his company and the Virginia Fire Department. He was a protégé of Tom Peasley and others of the rough element of Virginia City, and was not above using his fists when the occasion arose. William Sheppard was a miner, and a member of Washoe Engine Company No. 4, with a reputation similar to that of Ballou. So, it was not surprising that when they met on the boardwalk in front of Pat Mulcahy's Capitol Saloon on March 2, 1866, one of them ended up dead.

Ben and Billy began bantering about something, and before the argument was concluded, Ballou landed a fist square on the miner's chin, then went into the saloon. Sheppard entered right after Ballou and the fight continued, with Sheppard drawing a derringer and firing a single round into Ballou's forehead, just over his left eye. "His boots were pulled off at once like Tom Peasley's, that he 'might not die in them' & he was laid on counter," Alf Doten wrote in his journals. "[He] had best of surgery, but remained perfectly insensible till 12 at night when he died." Sheppard was arrested immediately after the shooting and taken to the city jail. Ballou's death, on March 2, 1866, inside the engine house of Eagle Engine Company No. 3, stirred up the firemen of the town. The thirty-year-old native of New York had been a popular fireman, and Sheppard was equally popular among his peers. It was a difficult situation for all involved.

An inquest was held the next morning, with a verdict that Ballou had died from a single gunshot at the hands of Sheppard, who would be tried for murder. The next day, all of the fire companies of Virginia City and Gold Hill, including Sheppard's Washoe Engine Company No. 4, turned out for Ballou's funeral, with about 290 persons on foot, and several carriages in the procession. Ballou's body was shipped to San Francisco's Lone Mountain Cemetery, where he was buried near his old friend, Tom Peasley, in the firemen's section. Sheppard was tried for murder a week later. It took the jury only ten minutes to reach a verdict of innocent, but because of the large crowd which had assembled, Judge R. S. Mesick deemed it prudent to wait an hour before announcing it. When the verdict was finally made public, a large cheer went up, and the jury and about 200 of Sheppard's friends retired to the International Hotel for a celebration. A special benefit performance was held four days later at Piper's Opera House, to raise funds for Sheppard's court costs.[7]

If personal friendships among the firemen were strong, so, too, were the friendships between the various companies. In 1869, a delegation of members of Liberty Engine Company No. 1 traveled to lower Gold Hill to call upon the members of Lincoln Hose Company No. 3. "The boys were well received and handsomely treated," reported the *Gold Hill News*, "and in response some of the Lincoln boys last evening came up and paid the Liberty boys a similar visit. Putnam showed us sundry empty beverage bottles and crumbs of lunch at the engine house this morning, and says they had a jolly time."[8] It was not uncommon for one fire company to aid another in time of need. When Eagle Engine Company No. 3 was struggling to raise funds for the construction of a new engine house in 1872, Knickerbocker Engine Company No. 5 donated the entire proceeds of their Thanksgiving Ball, $479, to the Eagle building fund.[9]

Courtesies among the firemen could mean the simple invitation for one company to drill with another, or an elaborate invitation to join them at a fancy dress dinner and ball, picnic, or other celebration. A grand drill of the Gold Hill Fire Department was organized in 1874 by Yellow Jacket Engine Company No. 2, with personal invitations extended to Liberty Engine Company No. 1 and Lincoln Hose Company No. 3. Virtually every fireman in the department turned out for the drill, and afterwards retreated to one of the local saloons to hoist a few flagons of ale and cheer the performance of the fire companies and their apparatus.[10]

Parades were another occasion to extend a friendly invitation to neighboring fire companies. In 1896, Liberty Engine Company No. 1 voted to invite members of Silver City Hose Company No. 1 to march with them in the Fourth of July parade.[11] The next year, Liberty Engine Company No. 1 accepted an invitation from Curry Engine Company No. 2, of Carson City, to attend a formal uniform dance and supper.[12] In 1899, the Butt Enders of Gold Hill chartered a special train on the Virginia & Truckee Railroad, to attend the annual dance of Warren Engine Company No. 1, in Carson City.[13]

Helmet shields, firemen's belts, helmets, and photographs of firemen were often exchanged between the various fire companies as tokens of friendship. Grand tributes, executed by masters of penmanship, were prized tokens of friendship from one company to another. Hanging in the museum of Warren Engine Company No. 1 in Carson City is a large scroll presented by the firemen of Liberty Engine Company No. 1 in 1876. The *Gold Hill News* described the resolution from its fire company just prior to the presentation, in August of the centennial year:

> The Centennial Greeting which is to be presented by Liberty Engine Company No. 1 of Gold Hill to Warren Engine and Hose Company

Dance card from the 1893 anniversary ball of Warren Engine Company No. 1 in Carson City. Social events played an important part in a fireman's life. (Author's collection)

No. 1 of Carson, is the neatest thing in that line we have had an opportunity of seeing for a long time. It consists of a set of resolutions passed by the Gold Hill firemen on the 12th of July, 1876, expressing their grateful acknowledgements for the generous reception tendered them by the Warrens upon the occasion of the Centennial celebration last Fourth of July. The resolutions are elegantly and elaborately engrossed upon a sheet of Bristol board, and surrounded by a beautiful frame, twenty-six by thirty-six inches inside, and from ten to fifteen inches wide. The frame was

> manufactured in San Francisco especially for its present use under order from the Liberty Company and is embellished by figures in relief representing engines and jumpers, crossed trumpets, crossed ladders and hooks, hydrants with a section of hose attached, torches, axes and pipes, and in fact, all the paraphernalia used by firemen. In each corner is a fireman's hat.
>
> On the sides in German text are written the mottoes of the companies. The Warrens say, "To Save Life We Risk Ours." The Liberty boys have proved themselves worthy of the words, "Faithful and Fearless."
>
> The engrossing was executed by Metcalf, of Virginia, and the work is the finest style imagineable. The very best feeling now exists between the two companies and such testimonials of friendship as this will tend to more strongly unite them.[14]

More than one hundred years later, those strong ties of friendship between the two companies still exist, as the companies repeatedly extend to each other traditional courtesies, with invitations to attend special dinners, fund raisers, and other events together. They also continue the tradition of the firemen's tournaments, by sharing the responsibility of organizing and running the annual Comstock Firemen's Muster each summer. The members of Warren Engine Company No. 1 and Liberty Engine Company No. 1 share a history of more than a century of fighting fires in their respective communities, often alongside each other.

Through the years, rivalries between the fire companies on the Comstock, though heated at times, were for the most part unbloodied. But in the early years of the Virginia Fire Department one rivalry was carried to the extreme.

Comstock firemen were tough men. Many of them came from the ranks of the New York Fire Department, where the "Bowery Boys" had taught them the virtues of street fighting with Bowie knife and club. Some had become expert at wielding spanner wrenches or firemen's trumpets during their days as members of one of the companies of the San Francisco Fire Department. Others had been soldiers, teamsters, or cowboys, or were just hard men who had learned to survive by striking first and hardest.[15] Despite their qualifications for street brawling, their pride in their equipment and respective companies, and their penchant for braggadocio, the rivalry between fire companies on the Comstock splashed blood upon the streets of Virginia City only once. It left a stain so permanent and vivid that it was never to happen again.

The rivalry started after Virginia Engine Company No. 1 and Nevada Hook and Ladder Company No. 1 were formed in 1861. Both companies were made up of men who had initially formed the engine company. Although designated by two different names and functions, they acted as if they were but one company. When Virginia City's second engine company was formed, under the leadership of tinsmith Jacob Young, Jr., the trouble began.

Young America Engine Company No. 2 sought an engine rated better than that of Virginia Engine Company No. 1, and purchased the former Monumental Engine No. 6 of the San Francisco Fire Department. Even before its arrival, the new engine had sparked several saloon arguments over which company had the best engine, and who could get "first water" at a fire. Young America Engine Company was made up principally of working men who prided themselves on being "the boys," while Virginia Engine Company No. 1 and Nevada Hook and Ladder Company No. 1 prided themselves on being the "sports" of the town.

The next serious fire to strike Virginia City broke out in a carpenter shop in the rear of Patrick Lynch's saloon, near the corner of C and Taylor streets, on August 29, 1863. The burned district extended from Taylor Street to Sutton Avenue, north and south, and from A Street to B, and partly down C, east and west, covering virtually the entire business district of the town. The fire destroyed $700,000 worth of property and left one man dead—from a gunshot.[16]

The rivalry between the first two engine companies had continued, even though two other engine companies had joined the fire department. During the heat of the fire, firemen from the rival engine companies and the hook and ladder company turned their fighting efforts from the blaze and towards each other, leaving Eagle Engine Company No. 3 and Washoe Engine Company No. 4 to contend with the fire. Precipitating the fight was Edward Richardson, an expelled member of Young America Engine Company No. 2, whose mouth got the best of him and involved virtually everyone in the three rival companies in the fight. Brickbats were sent sailing through the air and spanner wrenches were used as clubs. As buildings burned behind them, the firemen of the city battled it out. More men were injured in the fight than from fighting the fire, and many received medical attention for serious cuts and contusions received from flying objects.[17]

The fight had nearly reached its peak when, from the center of the mob, a shot rang out from the pistol of John P. Cullen, second assistant chief of the Virginia Fire Department, and a member of Young America

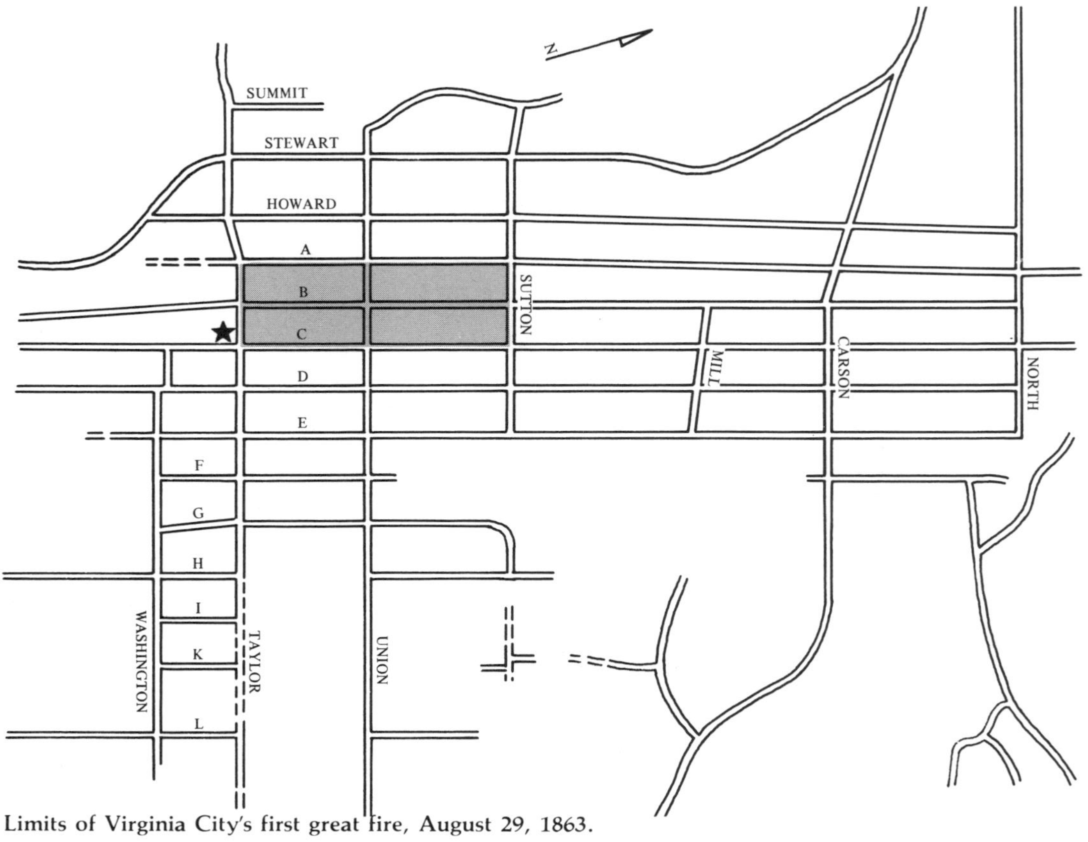

Limits of Virginia City's first great fire, August 29, 1863.

Engine Company No. 2. Richardson, who was in the center of the melee, slumped to the ground mortally wounded. Every fireman in each of the companies immediately grasped the seriousness of the situation. As soon as the fire was extinguished, the members of Young America Engine Company enveloped Cullen in a protective wave of firemen, and moved towards the sanctuary of No. 2's engine house on south C Street.

An angry mob gathered and began to follow the firemen towards the engine house. Most of Richardson's friends were among the crowd, including several antagonists from the rival engine and hook and ladder companies. As the firemen retreated to the front of their engine house, shouts and threats were made. A promise to dismantle No. 2's house stick by stick became a battle cry for the angry mob. Young America's firemen, although not eager to shed blood, were more than ready to defend themselves. They loaded their small signal cannon with grapeshot and cannister, promising a river of blood in the streets of Virginia should the crowd advance upon them.

Billy Warnock, a member of Washoe Engine Company No. 4, prevented violence with a moving speech warning both sides against action that would only lead to disaster. He cautioned Richardson's friends that an advance upon a loaded cannon with forty or fifty armed men behind it would be foolhardy at best. The crowd took the hint and withdrew, but only after promises had been made that Cullen would be brought to justice.[18]

Cullen was arrested for murder and placed in the city jail. He appeared before Judge Locke in November. The trial was not a long one, but the jury could not agree upon a verdict. Nine members favored acquittal, two recommended manslaughter, and one insisted upon murder in the first degree.[19] Cullen was a confectioner and a popular fireman. When the facts of the shooting and the fight between the rival fire companies became known, support for the thin, frail-looking fireman increased. The jury was discharged and a new trial was ordered. Cullen was released on $10,000 bail while he awaited the new trial.[20]

The firemen of Young America Engine Company No. 2 displayed their support of the accused Cullen in February of 1864, by nominating him for re-election as first assistant chief of the fire department. Cullen was not on hand for the election, however. On March 22, 1864, Judge North sentenced Cullen to two and one-half years in the Nevada State Prison in Carson City, after a jury convicted him of the murder of the wayward Richardson. A new wave of support for Cullen swept over the

Comstock, and townspeople were urged to sign a petition seeking a light sentence for the convicted fireman. According to the *Gold Hill News*:

> The judge said that while he fully appreciated the feeling of humanity which prompted so many citizens to petition him in favor of the prisoner, and although he honored them for this humanity, yet he could not consistently, with what he felt to be his duty, make the sentence lighter than what he had done. There was an immense crowd of persons in the court at the time, who appeared somewhat surprised at the severity of the sentence, as did also the prisoner, who with his friends expected imprisonment for a year as the very limit of his sentence.[21]

Cullen served his sentence and was released at the end of his term. He returned to Virginia City and again served in Young America Engine Company No. 2, holding offices in the company and running with members of the company's hose cart race team in 1869.[22] Cullen continued to serve in the fire department, but at each fire the excitement of men shouting orders and running about must have reminded him again of that August day in 1863, when a fire alarm turned into a riot and ended in tragedy for the town, himself, the fire department, and Edward Richardson.

Cullen may have remained active in the fire department until it was disbanded after the Great Fire of 1875. Eventually, according to Cullen family records, he migrated to Pioche, and later to Bodie, California, where he was proprietor of the Bank Exchange Saloon. In 1878 he helped organize Bodie's first fire company, Babcock Engine Company No. 1.

Chief Tom Peasley: A Tough Man

TOM PEASLEY was a tough character with leadership qualities that made men from all ranks of society follow him. He was a man with strong convictions, which he would defend with fists, knife, or gun. Peasley was one of the organizers of the first engine company in the state, Virginia Engine Company No. 1, and he also worked to organize Nevada Hook and Ladder Company No. 1.[1]

Tom Peasley was one of the town "toughs," and frequented the town's many saloons and bawdy houses. He was a staunch Union supporter at a time when secession sentiment was growing throughout Nevada, including the Comstock. His patriotism was more than rhetoric, and it burst forth when secessionists interrupted a Union Army recruiting drive on the streets of Virginia City by breaking the drum of one of the marchers and shouting loud support for the new Confederate government. Peasley and several other members of the fire department used their fists to quiet the uprising, and then formed an escort to march the young Union lieutenant and the two boys with drums to the courthouse, where speeches were given and many new Union recruits were signed up.[2] It was this kind of prompt and direct action that earned Peasley a reputation for no-nonsense handling of problems and helped win him appointment as sergeant-at-arms of the Nevada Legislature, as city policeman, and later as city jailer.[3]

The tough saloon keeper was known to keep company with Julia Bulette, one of the town's prostitutes. Their relationship may have led to her honorary membership in Virginia Engine Company No. 1 at the time Peasley was a member of that company.[4]

His leadership qualities and tough reputation helped win his election as the first chief engineer of the Virginia Fire Department, an office he held from 1861 through 1862. At the end of his term as chief, Virginia Engine Company No. 1 voted to endorse his rival for the post, Peter Larkin, who defeated Peasley to become the second chief of the fire department.[5] Peasley's pride suffered at the insult delivered to him by his own company, as well as his defeat by the firemen of the department as a whole. As a result, Peasley transferred his membership from Virginia Engine Company to Eagle Engine Company No. 3, which had been organized by his brother, Andrew. Later, both men would hold the offices of President and foreman of that company.[6]

Peasley's reputation as a town tough was well earned. In 1864, he was indicted for the murder of Sugar Foot Jack, who had sworn to kill Peasley the minute he laid eyes on him. Peasley wasn't one to wait for trouble to find him, however. Instead of waiting for Sugar Foot Jack to make the first move, Peasley hunted throughout the town for him and shot him where he found him. The popularity of the tough fireman, however, worked in his favor. He was acquitted by jury members who did not even leave their seats for deliberation.[7]

Later that year, Peasley became involved in an argument with Martin V. Barnhart, a member of Warren Engine Company No. 1, in a Carson City hotel. A shoot-out followed, and Peasley put two slugs into Barnhart, who, the newspapers reported, "was slightly prevented from returning fire."[8] Several months later the two met again, in the Ormsby

Thomas Peasley, Virginia City's first fire chief. (Special Collections, University of Nevada, Reno, Library)

House in Carson City. Barnhart's wounds had healed, but not his anger, and shots were again fired. Barnhart missed, but Peasley scored a hit. On February 2, 1866, the two combatants met again in the Corner Bar at the Ormsby House. Several of the "boys" were whooping it up at the bar when Barnhart walked in and strutted up to Peasley, poking a cocked revolver into his chest. Barnhart accused Peasley of cowardice for refusing to fight him on another occasion, called him some hard names, then stated he was going to shoot Peasley, which he did.

The first shot hit Peasley in the chest. He fell forward, grabbing Barnhart's gun from him. Barnhart grabbed it back and began clubbing the Virginia City fireman over the head with it. Peasley backed away, drew his pistol, and pumped five shots into Barnhart, who staggered out the door, tearing it off its hinges as he passed. The mortally wounded Peasley lunged forward and, with virtually his last breath, fired a final shot into Barnhart, who fell dead in the street. As he lay dying on the saloon floor, Virginia City's first fire chief called his friends to remove his boots, so he might not die with his boots on.[9]

The next day, Alf Doten wrote in his diary:

> At 5 p.m. the fire bell struck the alarm & all the fire laddies met at No. 3's house—flags all halfmast & No. 3's house draped in mourning—Tom was one of 3's boys—at 5 when the bell sounded, the boys all went to 3's house & from there to [the] Divide & escorted Tom's remains into town—in hearse—went to 3's house where he was laid out in state.[10]

Peasley's funeral was held two days later, from the hall of Eagle Engine Company No. 3. The fire departments of Virginia City and Gold Hill were represented by huge delegations from the various companies, as well as a delegation from Warren Engine Company No. 1, all wearing their dress uniforms. Doten noted that the procession numbered more than four hundred, and speculated that had the streets not been so muddy there probably would have been more than one thousand mourners.

Twelve members of Eagle Engine Company No. 3 were the pallbearers, dressed in red shirts, black pants, and the traditional fireman's leather helmet. An escort of twelve mounted members of the company surrounded the hearse, and the coffin was draped with the American flag. Unmounted mourners marched along the city's boardwalks rather than slosh through the muddy bog in the streets. They escorted the casket out to the north end of town, where it was taken to Reno for shipment to San Francisco. Peasley was buried there, in the firemen's section of Lone Mountain Cemetery, because of his affiliation with one of the pioneer San Francisco fire companies. Many years later, the cemetery was moved to another site, and the location of Tom Peasley's grave was lost.[11]

Martin V. Barnhart was buried in Carson City's Lone Mountain Cemetery, in ceremonies less elaborate than those accorded Peasley. Barnhart's grave is marked by a sandstone obelisk that provides no hint of the drama of his demise.

VIII. Man the Brakes!

Whether it was an Agnew end-stroke, a Button & Blake, a Jeffers side-arm or a Hunneman "tub," it required anywhere from twenty to eighty men, and considerable strength and endurance.

"Man the brakes!" the foreman would yell to his men, who were already panting from the run to the fire. "Ready now, bring her up easy. Okay now, up, down, up, down. One, two. One, two . . . " It was hard work, pumping the fire engines of the Comstock, work that required dedication and physical conditioning. But the firemen of the Comstock were proud of their engines. They worked them hard at fires and contests, to prove they were the best hand pumpers in the department, or anywhere.

The hand pumpers were physical symbols of the engine companies of the old volunteer fire department. Most of the engines in Virginia City were of different makes, gaily painted and decorated, reflecting the different tastes and slogans of the various fire companies. An engine might be surmounted by a gilt eagle bearing the number of the company in its beak, fancy iron scrollwork, and huge bells that rang continuously as it lumbered down the rough dirt streets of the town, pulled by ten to sixty men. Without exception, the engines bore the name, and often the motto, of the engine company. "Prompt to Act When Danger Calls" adorned a shield mounted in the scrollwork above the engine of Knickerbocker Engine Company No. 5.[1] "Faithful and Fearless" appeared in fancy gold lettering upon the hand engine of Liberty Engine Company No. 1 in Gold Hill.[2]

When the alarm bells or steam whistles sounded, signalling a fire in some section of the city, the volunteer firemen scurried to their respective engine houses. They hauled the hand pumpers to the conflagration, extended and manned the brakes, and, if they were the first to arrive, announced their triumph with the shout, "First water!"

The engines were usually trimmed with gleaming brass and nickel, which the firemen kept bright and shiny to show their pride in the "masheen." In addition to leather buckets, axes, and hooks, the machines also had fine lamps with cut glass, often colored, bearing the name, number, or motto of the company.

Virginia Engine Company No. 1 brought the first fire engine to Nevada. It was one of the most powerful fire engines on the West Coast in 1861, with 9¼-inch cylinders and 7½-inch stroke. It was built in Waterford, New York, by the Button & Blake Company, specifically for Virginia Engine Company No. 1.[3] Manned by up to forty-four men, the little engine served Virginia City for more than ten years. Its company shunned steam engines and fancy chemical extinguishers. No. 1 was a favorite in parades and squirting contests, and could always be counted on to do good service at a fire. It was a strong hand pumper, as the *Gold Hill News* noted in 1865:

> The engine took water from the cistern on North C street, near the Music Hall, and playing through 50 feet of hose and an inch nozzle, threw a horizontal stream of one-hundred and seventy-four feet, as measured by Chief Larkin. The engine has done better by a few feet, but the work yesterday shows a very efficient company, and under the Foremanship of the consumptive Brown it is hard to beat.[4]

As with all Comstock fire engines, No. 1 was sometimes subjected to hard treatment. In running to the Star House fire on B Street in 1865, Virginia Engine No. 1 was smashed against a building on the corner of C and Sutton streets, disabling at least one wheel. At another fire in 1866, the engine was disabled when some rascal punched a large hole in the air dome.[5] Some time later, Virginia Engine No. 1 had an interesting, if somewhat unorthodox use at one of the rowdier brothels in Virginia City's red-light district on D Street. Alf Doten wrote, probably with a slight smile, "Some fellows took No. 1's engine about 4 o'clock this morning & washed out old Cad Thompson's whore house—gave her hell—created quite a consternation among the law & order portion of the community."[6]

No. 1 was a faithful servant and fought many fires. During the Great

Fire of 1875, No. 1's brakes were continuously manned, often by persons who were not members of the volunteer company. The rough working that the pumper received during this fire resulted in severe damage to the leather cups in the cylinders of the engine. After the fire, when the fate of the volunteer fire department was in doubt, and while construction to replace the burned engine house of Virginia Engine Company No. 1 was underway, an appeal was made to the Board of Aldermen for funds to repair the engine. This request fell on deaf ears, as the board held steadfast to its policy of refusing funds for major repairs to any of the companies of the volunteer fire department.[7]

Eventually, all of the company's property except the hand engine was sold to the city. In 1877, the remaining members of Virginia Engine Company No. 1 voted to donate the hand engine to the Virginia Exempt Firemen's Association, to be kept in the growing museum of fire relics beneath the hall of the Exempts on C Street. The Exempts accepted the engine with gratitude, expressed in resolutions printed in the local newspapers. The engine was given a place of honor among the relics of the other companies of the old volunteer fire department. Some of the fire companies sold their old engines in favor of new ones, so very few of the old hand pumpers remained on the Comstock for very long. After Young America Engine Company No. 2 introduced a steam fire engine to Nevada in 1872, it spelled the end for most of the hand pumpers in the Virginia Fire Department, as well as in the fire departments of other towns and cities in the state. At various times, however, it was stored in the quarters of the new Virginia Paid Fire Department.[8]

The Exempts proudly hauled their old engine in parades along C Street for many years, and resisted efforts to sell the engine. When the ranks of the Exempts began to shrink and the organization could no longer muster enough members to haul the state's first fire engine, it was loaned to the volunteers of Divide Hose Company No. 2 for parades and celebrations.[9] Despite resolutions that the engine "forever remain" in Virginia City, the Exempts yielded to the demands of a tight economy, and accepted an offer to purchase the engine in 1896. Virginia Engine No. 1 was sold to the Gardnerville Fire Department, where it served for many more years before it was sold in the 1930s to Parker Lyon. He displayed the engine in his Pony Express Museum in Arcadia, California, and occasionally loaned the engine to movie studios. The current location of No. 1 is unknown.[10]

Young America Engine Company No. 2 was responsible for bringing to Virginia City the first rivalry in the fire department, as well as the first of what would be several pieces of equipment formerly serving the San Francisco Fire Department. The men of No. 2 wanted an engine larger than that of Virginia Engine Company No. 1, and went to San Francisco to find one. The engine they brought back to Virginia City was the former Monumental Engine No. 6, and was known in San Francisco as "Big Six." It cost $6,000 in 1862.[11]

Young America's new hand pump was a powerful end-stroke, with 10-inch cylinders and 9½-inch stroke, capable of producing five streams of water at a time. "Young America No. 2, threw yesterday, at their drill, a side stream of water through a 1⅛-inch nozzle a distance of two hundred and nineteen feet—this is the biggest throw, of record, on the coast," the *Gold Hill News* reported in 1863.[12] In 1865, the newspaper reported No. 2 "threw a stream far above the flag staff on the International Hotel, and gave proof of a first class machine in excellent order."[13]

Like Virginia Engine No. 1, Young America's hand pumper also suffered its share of mishaps. While running to a fire in 1867, No. 2 turned down Taylor Street and smashed into a house. Only two men were on the tongue and no one had taken a position at the brake. As the ponderous engine turned the corner, the men on the tongue found it unmanageable, and let go. The engine tore off about half the side of the house, and damaged its own brakes and other running gear.[14]

The heavy old hand pumper served Young America Engine Company No. 2 until 1872, when the company purchased a new steam fire engine, the first in the state. The hand engine was paraded one last time during the Fourth of July celebration, then was put up for sale. The *Territorial Enterprise* later reported on its history:

> Young America Engine Company No. 2, having purchased a steamer, yesterday disposed of their hand engine. It goes to Corvallis, Oregon, and the company received for it $2,000 gold coin. The engine is a fine one, and although the Young Americas have a steamer, they disliked very much to part with the old machine, and tears were to be seen in the eyes of more than one fireman as they rolled her out to be sent to Oregon. The engine was first bought in 1853 by Monumental Engine Company, San Francisco, where she did good service for ten years, when she was brought to this city. Here she has always done excellent work. Nearly a hundred men can be worked on her brakes, and she throws, when properly manned, three good streams. The engine passed out of town last evening behind a team of four horses. At Steamboat Springs it will

be put aboard the cars and thus shipped to San Francisco, where the Oregonians take charge of it and take it home.[15]

Eagle Engine Company No. 3 brought another San Francisco engine to Virginia City, the former Vigilant No. 9, a Jeffers side-arm pumper that cost $3,700. The Eagles' new engine weighed 4,000 pounds, "a pretty good load for the boys to haul up some of the steep grades of our city," the *Virginia Evening Bulletin* noted upon the arrival of the engine, in 1863.[16] At a test in 1865, the Eagle engine distinguished itself, the *Gold Hill News* reported, by "throwing a stream through an open butt . . . a distance of sixty-six feet, which was the farthest attained by any engine on the coast to that time."[17] In an 1870 contest, the Eagle engine threw a stream 210 feet, 9 inches, through a single length of hose and a ⅞-inch nozzle.[18]

The weight of the engine did create problems for the firemen who had to haul it up and down the steep streets of Virginia City. But No. 3's members were equal to the problem. In 1865, they fashioned a rough lock or iron shoe to fasten to the wheel of the engine, to help in braking when taking the heavy machine downhill. The device, the local newspapers said, "will perhaps prevent broken limbs or mere loss of life."[19]

No. 3 was another reliable engine which turned out at every sounding of the fire alarm bells. It was strategically placed during the Great Fire of 1875, and was credited with saving at least one residence from destruction, by putting out each fire that ignited on the house during the conflagration. It was also credited with checking the spread of the fire in its district, and thus saving much property which might otherwise have been turned to ash.[20] When Eagle Engine Company No. 3 disbanded in disgust, amid sharp criticism of the volunteer fire department following the Great Fire of 1875, the members voted to sell all of the property of the company, including the old hand engine. They divided the proceeds among themselves, and old No. 3 passed from view quickly and quietly.[21]

Washoe Engine Company No. 4 wanted nothing to do with a used fire engine, so the company ordered a new Button & Blake hand engine be built especially for them, at a cost of $2,800. The new engine, which had 9½-inch cylinders and 7½-inch stroke, was lighter than the other engines in the Virginia Fire Department. It was set on springs, which the press called "a marked improvement for this rough country." The engine also featured Button & Blake's new patent air chamber, and was, at the time, the only engine on the West Coast with that improvement.[22]

The side-arm Jeffers hand pumper, purchased by Eagle Engine Company No. 3 in 1863, was one of several former San Francisco engines to see service on the Comstock. Photo was taken on B Street, Virginia City. (Special Collections, University of Nevada, Reno, Library)

The first trial of the new engine took place four days after its arrival, in November, 1863, at a cistern on C Street, just south of Union Street. Hose was attached, and the firemen manned the brakes, putting two streams over the flagstaff on the International Hotel. On another test, with a single hose and a 1¼-inch nozzle, the engine put water at least 30 feet higher than the flagpole. The test was hailed as "the finest throwing ever accomplished on this side of the mountains, and the boys are justly proud of their machine."[23]

Later that month, No. 4 met with Young America Engine Company No. 2, for a test of the two engines. Washoe No. 4 sucked the cistern dry in "washing" No. 2's firemen, and the feat of the new fire engine was gracefully acknowledged by No. 2 sending a foxtail up to No. 4.[24] On another occasion, the Washoe hand engine threw a stream 201 feet, through 100 feet of hose, using a ⅞-inch nozzle, spray not included. "This work is hard to be beaten," the *Gold Hill News* declared, "and the boys, of course, feel proud of their achievement. Four's tub has a first rate crowd of boys, and is a component part of a Fire Department that we believe has no equal in the world."[25]

When it came to a dispute over who put "first water" on a fire, Washoe Engine Company No. 4 stood on its past achievements, and its tall squirting record. So strong was the reputation of the fire company that the *Gold Hill News* settled a dispute in its favor by stating, "The Virginia papers have been disputing the last two or three days about which company got 'first water' on the recent big Chinatown fire, No. 3's or No. 4's. It is finally settled in favor of No. 4's. This is correct. Whenever the matter of 'first water' is in doubt, it is always safe to give it to Four's." Despite the very strong record of the engine, the newspaper failed to mention the fact that its reporter, Alf Doten, was also a member of Washoe Engine Company No. 4, a fact which no doubt had some bearing on the angle of the story and its conclusion.[26]

In 1872, following the lead of Young America Engine Company No. 2, Washoe Engine Company No. 4 voted to purchase a steam fire engine. With delivery of the steamer, the old hand pumper took a back seat, but remained in the engine house. Doten's report on the new engine perhaps reveals the reluctance of other members as well: "We don't know whether to approve of this change or not, for the machine they now have is famous for its lightness, always getting first water. It is the best hand machine we have ever seen for a mountain locality, and she should not be allowed to leave." In any event, the engine remained in service in

Virginia City until the old volunteer fire department disbanded after the Great Fire of 1875.[27]

The decision to retain the engine was a wise one. The new steamer of No. 4 was out of service for repairs when the Great Fire broke out, and the reliable old Button & Blake hand engine was hauled out to fight the blaze. After the fire, when the volunteer companies were disbanding, the old hand engine was sold for $800 to Reno's Engine Company No. 1. It remained in service there until it was sold again in 1887 to Bellevue, Idaho.[28]

The city's fifth fire engine was brought to the Comstock in 1864 by Knickerbocker Engine Company No. 5. The Hunneman engine was described by the *Territorial Enterprise* as "small and light." In a contest with Washoe No. 4, the "Knicks" squirted 180 feet, showing the capabilities of the engine to residents of the city's First Ward, where it was to be housed.[29]

Five years later, the engine company voted to purchase a new engine. They entered into negotiations with the Marysville Fire Department in California, for the purchase of "Eureka Engine No. 1," which had been built for Marysville in 1856, at a cost of $4,500. The new engine was a Button & Blake, weighing 3,500 pounds, and of the same style as No. 4. After No. 5's new engine arrived in 1870, the *Gold Hill News* reported:

> Knickerbocker Engine Company No. 5, Virginia, had their new machine out yesterday for another trial of her squirting powers, and Washoe Engine No. 4 was brought out for a little match with her in that respect. Both took water at the cistern on C Street, just north of Union, and each used seven-eighth [inch] nozzles. No. 5 claims 202 feet, horizontal throwing, and No. 4, 201 feet. Both did well. We understand that the cylinders of No. 5's machine, however, are a size larger than those of No. 4.

The old Hunneman engine which had first served the "Knicks" in Virginia City was sold to Dayton's Knickerbocker Engine Company No. 1, and was that city's first engine. In reporting the transfer of the engine, the press announced, "Old Knickerbocker's an excellent machine; she has done first rate service and is fully able to do a good deal more. The Daytonites should have invested in a fire engine long ago."[30] The two engines were destined to cross paths once again, however.

Knickerbocker Engine Company No. 5 disbanded after the Great Fire of 1875, and sold all of its equipment to the city, including the Button &

This Hunneman hand pumper was purchased by Knickerbocker Engine Company No. 5 and brought to Virginia City in 1864. The engine was sold to Dayton in 1870 and was purchased by Yerington in 1894, when this photo was taken. (Author's collection)

Blake hand engine. This was purchased by the Sutro Fire Department's Knickerbocker Engine Company No. 1, in 1879, for $200. The old engine, which featured a massive brass air chamber with "Our Own" in fancy enamel lettering, remained in service in Sutro until 1894, when it was sold to the Dayton Fire Department. Dayton's old Hunneman Knickerbocker engine had fallen into disrepair, and was sold to Yerington, where it finally passed into history. In 1936, the old Button & Blake was sold to Parker Lyon, who moved it to his Pony Express Museum in Arcadia, California. In 1955, gaming executive and automobile collector William Harrah purchased the Pony Express Museum and moved it to Reno for display. The old hand pumper, however, was stored and not displayed. On December 24, 1981, its existence was discovered by members of the Comstock Firemen's Museum, who endeavored to raise funds for the purchase of the old engine that had once protected Virginia

The only known view of the 1856 Button & Blake hand pumper used in Virginia City by Knickerbocker Engine Company No. 5. The engine was built for Marysville, California, in 1856, was brought to Virginia City in 1870, and was sold to Sutro in 1879 (photo was taken while engine belonged to Sutro). The engine was sold to Dayton in 1894, and then to W. Parker Lyon's Pony Express Museum in 1936. The engine was sold, with the rest of the museum, to William F. Harrah in 1955 and was purchased at auction in 1982 by the Comstock Fireman's Museum. (Comstock Firemen's Museum)

City. On January 31, 1982, the engine, which had seen service in Virginia City, Sutro, and Dayton, was purchased at auction for $7,500 and returned to the Comstock. It has been restored to working order and is today displayed in the Comstock Firemen's Museum in Virginia City.[31]

Another hand engine appeared on the Comstock with the organization of Confidence Engine Company No. 6, on the Divide, in 1865. As the *Gold Hill News* told its readers:

> During the past week Mr. D. Crosby has been in California, and while at Folsom purchased an engine for Confidence Company No. 6. The engine was purchased of Mr. Fred Holsinger, for the sum of $1,800, and is the same pattern and capacity as Knickerbocker No. 5. Mr. C also purchased a hose cart and 400 feet of hose, the whole costing $2,800, and being an excellent bargain.[32]

When the new Hunneman "tub" arrived on the Divide, it was immediately dismantled, thoroughly cleaned, and then set up. On her first trial, the engine threw a stream 174 feet through 100 feet of hose and a 1-inch nozzle. The pistons had not been packed correctly, though, and the press eagerly awaited another trial of the engine's capabilities. A week later, the *Gold Hill News* reported, No. 6 threw a stream 196 feet on trial drill with Knickerbocker Engine No. 5.[33]

In 1873, long after the company had been renamed Monumental Engine Company No. 6, it purchased a steam fire engine to serve the Divide, Gold Hill, and Virginia City.[34] The old hand engine of the company was retained, however, and it served on the Divide alongside the steamer until the company disbanded. In 1885, the hand pumper was sold to Ketchum, Idaho, and shipped out on the Virginia & Truckee Railroad.[35]

Liberty Engine Company No. 1 was the only company in the Gold Hill Fire Department ever to own a hand engine. After formation of the company as a hose company in 1863, the need for a hand engine became apparent. Several fires had wreaked havoc despite the town's gravity flow hydrant system and three fine hose companies. In 1868, Liberty Engine Company No. 1 brought another San Francisco fire engine to the Comstock when it purchased the former Howard Engine No. 3 for the sum of $1,100. "Although somewhat ancient," Virginia City's *Territorial Enterprise* said, "the engine is said to be a whale at 'skwirting,' and the boys down at Slippery Gulch threaten shortly to send a challenge to this city for a trial with our best machine."[36]

The heavy engine, with two sets of cylinders, gave a demonstration for

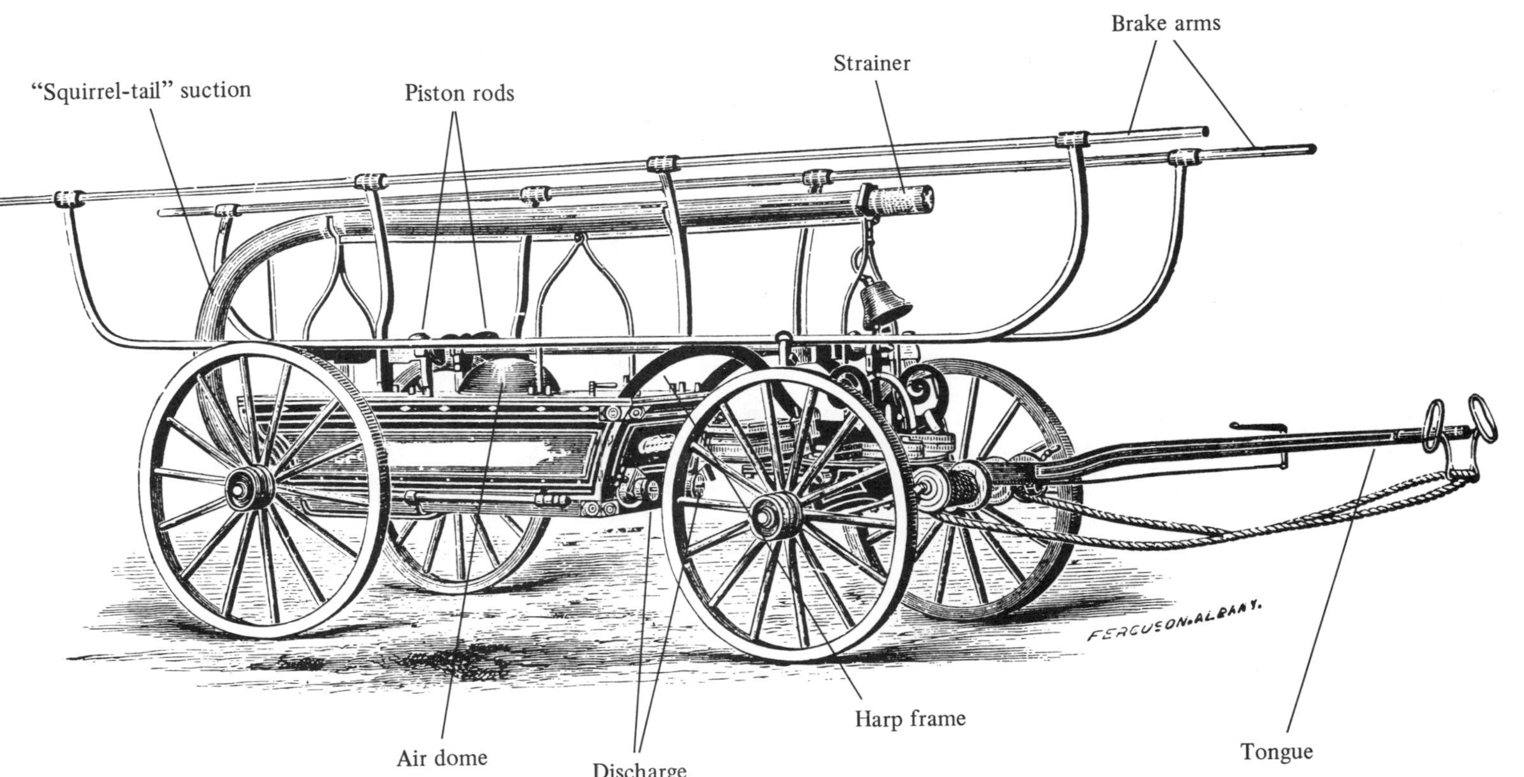

Button hand pumper, circa 1870, typical of the engines used on the Comstock.

the town shortly after its arrival. The *Gold Hill News* reported the engine "threw some very good streams, demonstrating how perfectly efficacious this engine would be in case of a fire where it could not be reached by means of hose from the town hydrant."[37] The Gold Hill firemen were proud of their engine, and when they were invited to attend the 1869 Railroad Celebration in Sacramento, they voted to haul the engine over the mountains on the railroad. "Old Howard No. 3, formerly of San Francisco, but now Liberty No. 1 of Gold Hill, was treated to a whole cartload of floral wreaths and trimmings as soon as she was taken from the cars, and as an old acquaintance among the old firemen, she was regarded with feelings of the highest regard and veneration all the time she was in town," the *Gold Hill News* reported upon return of the delegation from Sacramento.[38]

Although the engine performed reliably, it was too heavy, and its squirting capabilities were too small for the needs of Liberty Engine Company No. 1. The firemen complained that the wheels of the engine were too small, and that it sat entirely too near the ground for the rough, hilly countryside of Gold Hill, although they maintained that in a town where streets were somewhat level, it would be a perfect engine. In 1872, the complaints finally convinced the company to purchase a new engine. The company offered the old engine for sale, complete with hose, pipes, nozzles, tools, and other equipment, after it had been politely refused by Lincoln Hose Company No. 3, in lower Gold Hill.[39]

On March 24, 1873, Liberty Engine Company No. 1 took possession of its new Button & Blake squirrel-tail pumper, which was graphically described by the *Gold Hill News*:

> She cost, when laid down in Gold Hill, the sum of $3,200. The engine has all the modern improvements and is one of the finest of its class we have ever seen. The box is solid mahogany, inlaid with laurel wood stars, with gilt paneling and scroll work emblazoned on both sides in gold and vermillion. All of the iron and brass work is highly polished. The engine has two magnificent silver-plated signal lamps, the sides of which are red, white and blue glass. The front glass, which is blue, is decorated with a beautiful representation of the Goddess of Liberty cut in the glass. On the side glass, which is red, "Liberty 1" is neatly cut. On the suction sleeve is inscribed "Faithful and Fearless." The suction sleeve will be surmounted by a statue of the Goddess of Liberty, one foot in height. The wheels are 49 inches in diameter forward and 51 aft. A patent brake, which is attached to the hind wheels, is worked by

Liberty Engine Company's Button & Blake engine was the last old hand pumper to leave the Comstock. It was sold in 1910, for $450. The engine had a Honduras mahogany box with laurel inlays, featured a statue of Liberty atop the suction holder, and sported two fancy running lights. (Bob Dockery)

> both foot and lever, and is capable of stopping the engine instantly, even when going down the steepest grade. The weight of the engine is 3,500 pounds. The company expects to have her ready for inspection by the public in a few days.[40]

Liberty members spent the next few days polishing the brass, cleaning the wood, and getting the engine in shape for its first test. "Notwithstanding the unpleasant state of the weather," the *Gold Hill News* stated, "the engine behaved very handsomely and gave the most perfect satisfaction." The hand pump proved its worth on the steep and narrow streets of Gold Hill. It could turn completely around in a very short space, and was maneuvered easily despite the rough, hilly terrain of the town. But as the hydrant system of Gold Hill was developed and more fire mains were laid around the town, with additional hydrants spaced in critical locations, the old hand engine became more of a decoration than a working piece of equipment. It was hauled out only for occasional drills, demonstrations, and the Fourth of July parade. With strong pressure in the hydrant system, the town's firemen came to rely more upon the hose carriages and carts of the fire department than the hand pumper.

In 1894, Bishop Creek, California, offered to purchase the engine and requested a price from Liberty Engine Company. The company replied that the engine and all its accoutrements would cost $1,800, but the price was apparently too high for there were no further negotiations.[41]

A formal photograph of the engine was taken by the company in 1910. One month after a photographer named Baker came from Virginia City to capture the engine for posterity in front of the engine house, Liberty Engine Company President Tim Stack proposed to sell the old engine through a broker in Brooklyn, New York. According to the minutes of the company, the deal was to include a sale price of "$550 loaded on a car here and that to include the $50 commission." The minutes of the next meeting, held September 22, 1910, tell their own story:

> Pres. Stack stated that the object of the meeting was to act on the communication from Lowell, Mass., whether they would accept $450 for the engine.
>
> It was moved and carried that the Engine Co. accept the offer.
>
> John Price said that we ought to wait for a larger attendance at a meeting before we sold the Engine.
>
> Pres. Stack said that this would be as large an attendance we would get and if any of the members had any fault to find it was their own fault for not attending the meetings.

> It was moved and carried that the money for the engine be sent to the Treasurer of the Engine Co. before the engine was moved out of the Hall.
>
> The Secretary was notified to find out how they wanted the engine shipped and all the particulars about the same.
>
> It was moved and carried that the Engine Co. give H. H. Easterbrook $50.00 commission for making the sale of the engine when we receive the money for the same.

Shortly thereafter, the old Button & Blake hand pumper was loaded on a flatcar at the Virginia & Truckee depot in Gold Hill, destined for its new home. The last hand engine on the Comstock – the last remnant of the old days when firemen manned the brakes and yelled, "first water" as they pumped for a strong stream – slowly moved out of town on its way to Massachusetts.[42]

IX. The Hooks

They were among the toughest and rowdiest men in the fire department and had the distinction of being members of the only hook and ladder company in Virginia City. The members of Nevada Hook and Ladder Company No. 1 were a special breed. They had one of the most difficult jobs in the fire department, yet were always on hand to fight fire. The company was formed in 1861, from the ranks of Virginia Engine Company No. 1, the town's first fire company, and had some sixty-three members by the end of its first year of existence. It had a fine hook and ladder truck, which had been built by Folsom & Hiller in San Francisco, at a cost of $2,500. The hand-drawn truck carried seventy-five feet of extension ladders, leather buckets, pike poles or hooks, axes, rope, and other tools.[1]

Although they were firemen, the primary job of the "Hooks," as they were affectionately known, was destruction. It was not uncommon to see the members of Nevada Hook and Ladder clambering over a building while one adjacent was burning. It was their job to help prevent the spread of fire from building to building by tearing down structures adjacent to fires with the energy and determination of ants moving a large twig. The Hooks were also first-class firemen, and were often responsible for taking care of a fire before any engine company could claim "first water," as they did in 1863 when a frame building on B Street caught fire. They were also experts at salvage, which in those days meant the removal of furniture and other personal belongings from buildings in danger of destruction, or already being consumed by fire.[2] When the home of Judge Ferris caught fire in 1864, Hook and Ladder Company No. 1 was praised by the *Gold Hill News* for its efforts, "Nevada Hook and Ladder can't be beat, and they worked last night like heroes, and assisted greatly in subduing the flames by tearing down fragments of the building and cutting away adjacent outhouses."[3]

A number of the most prominent and colorful citizens of the Comstock were affiliated with Nevada Hook and Ladder Company No. 1, among them John Van Buren Perry, city marshal and foreman of the Hooks. Perry was a huge man and a staunch supporter of the Union. He had served in the New York Fire Department, where he had learned not only the hazards of fighting fires, but also the tough methods of survival. He brought that knowledge with him to Virginia City and used it as needed. Perry was one of Tom Peasley's protégés, both having been charter members of Virginia Engine Company No. 1 when it was formed in 1861. When Nevada Hook and Ladder Company No. 1 was formed, Perry became its foreman.

Mark Twain recalled Perry's effectiveness as a leader, and his Union convictions, in a description of Virginia City's reaction to the news that Union troops, including a regiment of former New York firemen, had been soundly defeated by Confederate troops at the first Battle of Bull Run:

> The *Enterprise* was getting out an extra containing the particulars of the disastrous defeat, about noon, when Jack Perry, who was an old pressman and therefore a privileged character, dropped into the office and asked to see the proofs. The giant fireman read the dispatches all through, how our army had been routed and the New York Fire Zouaves cut to pieces, without saying a word, but with the tears streaming like two giant water courses down his cheeks. Then he got up, slid his pistol a little more to the front where it would be handier, and left the office remarking simply, "No damned secesh had better crow within my hearing today."
>
> He must have infected every member of the Fire Department with his feeling; for an hour later when the extra was issued and the Southern sympathizers began to exult, one would have thought the Battle of Bull Run had struck the Comstock, instead of a mere account of it. When the sun sank behind Mount Davidson that afternoon, it had looked for the last time on secession's domination in Nevada Territory. The rough element of the Union spirit had asserted itself, and the timid but more numerous part quickly rallied to its support. Thereafter, there was no open jubilation over Southern victories.[4]

A fine hook and ladder truck built by Folsom & Hiller, similar to this, was purchased by Nevada Hook and Ladder Company No. 1. After the Great Fire of 1875, the truck was purchased by the city, but seldom used.

On another occasion, Perry, joined by Fire Chief Tom Peasley and firemen Jack Williams and George Birdsall, came to the rescue of a Union army recruiting drive on the streets of Virginia City. A young lieutenant, hoping literally to drum up recruits, had hired two local boys to march through the streets drumming. The little procession had gone less than one hundred yards when a Southern sympathizer sprang from the sidewalks and demolished one of the drums. He was in the process of destroying the other when the young lieutenant knocked him down. Reinforcements from the fire department appeared on the scene to insure the safety of the procession, which then marched to City Hall, where a patriotic address was delivered by Charles Bryan, former justice of the California Supreme Court. On that day, with the assistance of Perry, Peasley, and the other firemen, the young lieutenant signed seventy-five recruits for the California regiments, because Nevada had not yet been given permission to organize military companies under its own flag.[5]

Their politics were always serious, but the Hooks had a sense of humor as well. When only two members beside the president bothered to appear at a regular meeting, those present decreed the other sixty-two members of the company expelled, then quietly adjourned the meeting, scheduling another regular business meeting the next month.

A very special member of Nevada Hook and Ladder Company No. 1 was Lizzie Williams, who was made an honorary member of the company for her dedication to the Hooks. Titles meant nothing to Lizzie, who shed her "honorary" title at fires to become another one of the "boys" fighting the fire. She was particularly visible at the fire in 1869, when an enterprising reporter wrote:

> At the fire in Virginia yesterday morning we noticed among the firemen working at the brakes, one woman—Lizzie Williams. She worked just as hard, effectively and steadily as anybody, and her presence and example certainly had an inspiring effect, for we never saw that gay little machine handled livelier, or to better advantage. We doubt whether in any other city on the coast there can be found a woman who would thus work so freely and heroically on such an occasion, and in such weather. Standing in the cold snow and freezing night air, amid a driving snowstorm, she worked with a courage and enthusiasm only to be found among firemen, steadily refusing the bottled stimulants which were occasionally passed around to revive the tiring energies of the men who were working with her at the brakes.[6]

The Hooks were a valuable part of the fire department and worked as hard at fires as those on the engines or hose carts. When Virginia City's

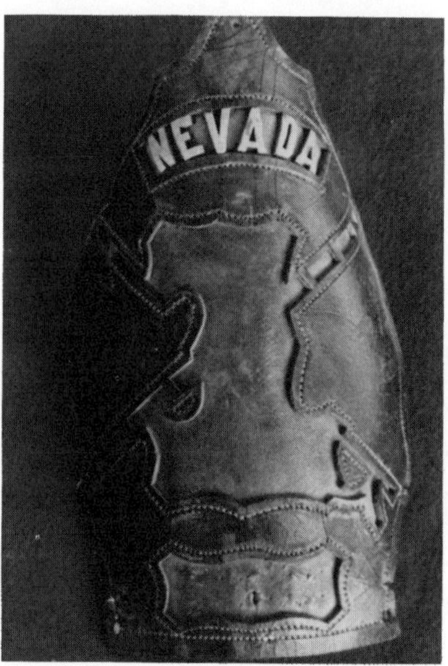

The only known helmet shield from Nevada Hook and Ladder Company No. 1 is in the Warren Engine Company No. 1 collection in Carson City. (Author's photo)

Chinatown caught fire in 1869, the Hooks were quickly on hand. Had they not gotten to work as fast as they did, more than the twenty buildings that were burned would have been destroyed. All of the engines in the city were quickly in place, putting water on the burning structures; but the members of Nevada Hook and Ladder were busy demolishing the adjacent dry, wooden buildings to prevent the spread of fire throughout Chinatown and possibly into the heart of the city's business district.

George Downey, foreman of the Hook and Ladder Company and chief of police, directed not only the efforts of his own company at that blaze, but also the efforts of an impromptu company of Chinese, who began tearing down everything in sight. The *Gold Hill News* recounted:

> The Chinamen rushed towards it from all quarters, and for the space of ten minutes we never saw any fire company equal them for either noise or energy. They attacked it on all sides, slashing away at it with wild desperation, some of them with their tails sticking straight out behind, like skillet handles, with excitement, while

> others rushed furiously, dashing big tins of water upon the flames. Under these vigorous circumstances the fire had to yield right speedily. Sides, roof, and all fell in together, with about a dozen Chinamen clinging fast and frantically thereto, like fighting bulldogs. By this time, however, there was no fire, it being completely deluged out. George Downey, who superintended the job, says he will match his China "Hooks" against any others on the coast.[7]

Downey's unofficial Hooks got their chance in 1870, when a firecracker explosion set fire to a wooden building in the center of Chinatown. Once again, Nevada Hook and Ladder Company No. 1 was on hand, and once again the China "Hooks" proved their worth. Due to the poor condition of the streets, only Knickerbocker Engine No. 5 managed to get to the scene, but it was hardly needed. The Chinese had taken control of the situation and put the main fire out before the hand pumper got on the fire ground. According to the next day's newspaper report:

> George Downey's long-tailed "Hooks" got afoul of that burning house and completely demolished it, tearing it down and dragging it to the four parts of the compass, before the flames could get a fair show at a successful operation. Those Chinamen worked like devils, swinging their axes and tearing away with claws, teeth and toenails in a perfect frenzy of desperation ludicrous to behold, all hands chatting and yelling like fury. A fire in Chinatown always creates more energy, noise and general fuss than anywhere else in creation.[8]

When the alarm of fire sounded on October 25, 1875, the Hooks were on hand to battle Virginia City's Great Fire. They fought alongside their brother firemen and, like others in the fire department, lost considerable equipment, including the firehouse on the west side of B Street, adjacent to the Miner's Union Hall. The following March, the company sadly reported that although its members were ready and willing to respond to any fire in the town, the truck was without ladders or hooks, all having been lost in the fire and not yet replaced.[9]

The company's lot was offered for sale for $1,000, and an offer was made to sell the old hand-drawn hook and ladder truck to the Board of Aldermen for $450. Although many thought the Hook and Ladder Company would be outfitted with a new firehouse and all the equipment it needed to be serviceable again, nothing was done. The old truck was finally taken to the Corporation House, where the new paid fire department was housed. The volunteers were disbanded, and the Hooks paid each member an equal share of the profits from the sale of the company property, including the old hook and ladder truck.[10]

Their hand-drawn ladder truck remained in Virginia City, refitted and ready for duty; but it was called out only for parades. The rest of the time it remained in storage, in the room the city had specially built for it in the new Corporation House. In 1900, Virginia City's second piece of fire equipment, the old hook and ladder truck, became the last piece of equipment from the old volunteer fire department to leave town, when it was sold to parties in Reno.[11]

X. *Jump Her Lively, Boys!*

The hose carts, carriages, and jumpers of the Comstock fire departments were often the most important pieces of equipment at a fire. They could carry up to one thousand feet of hose, and could easily be placed in a strategic location to battle a fire. They were the lifeline for the hand pumpers, but at times they were the only pieces of fire equipment at the scene of a blaze.

Most of Virginia City's engine companies organized hose companies to run with them during their first year of existence. Some were reorganized in the 1870s, however, with distinctive names. Virginia Engine Company No. 1 organized Rooster Hose Company No. 1, while the company that ran with Young America Engine Company No. 2 was known as Good Will Hose Company No. 2. Washoe Engine Company No. 4 christened its hose company Invincible Hose Company No. 4, while the men who ran with the hose cart of Knickerbocker Engine Company No. 5 discarded their earlier title, Our Own Hose Company No. 5, in favor of Neptune Hose Company No. 5. Monumental Engine Company No. 6, on the Divide, also had its own hose company, but its name has been lost.

In Gold Hill, the two engine companies were first organized as hose companies, later purchasing a hand engine and a steamer to become engine companies, although also retaining their own hose companies. Liberty Hose Company No. 1, which bought a hand pumper in 1868, also sponsored Junior Hose Company No. 1, made up of the youths of the town. This organization was similar to one organized by Eagle

Engine Company No. 3 in Virginia City, which sponsored the youths of Lincoln Hose Company No. 3.[1]

Until 1877, the Comstock's hose carriages were all hand drawn. They came in a variety of shapes, designs, and colors, with at least one brass bell, and sometimes as many as four. Several of the hose rigs were quite elaborate, with ornate ironwork surmounting the tool box in front of the hose reel, while others were plain in design, though highly functional. There were two-wheel jumpers, such as those designed by New York fireman David Hubbs, which were also known as "Hubbs' babies," in honor of the designer. These carts were designed to be pulled by a small number of men, or attached to the rear of an engine and pulled by more firemen. There were also large, four-wheel hose carriages, designed to carry a maximum of hose, nozzles, axes, and other tools. In either case, it required a minimum of eight men to pull the hose carts of the Comstock along the rough, steep streets of Gold Hill and Virginia City. Like the engines, the hose carts raced to the fire in a hell-bent-for-leather gallop to see which company could put "first water" on the fire and possibly redden the faces of the men on the engines.[2]

With the sound of the alarm bells came shouts of, "Jump her lively, boys!" as the foremen of the hose companies grabbed their helmets and firemen's trumpets and led the way to the fire. The carriages jumped along quickly and precariously over the rough streets, behind the men of the company, who were tightly clutching the tongue and tow ropes. The sound of the carriage bells clanging wildly over each bump issued a warning to those in the path of the speeding apparatus, but it was not uncommon for someone to become a casualty before the equipment ever reached the fire. The carriage of Invincible Hose Company moved so quickly to a fire in 1865 that it knocked down and ran over J. George, fracturing his skull, breaking his collarbone, injuring him so badly that he died the next day.[3]

The race to an 1876 fire in Gold Hill was equally perilous for the townspeople, as the three companies of the fire department vied for the honor of "first water." The *Gold Hill News* reported:

> The boys did jump around mighty lively at the fire last evening and nobody ever heard or saw so much noise and excitement as there was for a short time. Noise was no name for it. The big three-compartment whistle of the Imperial belched forth triple-tone thunder, the Yellow Jacket howled most fearfully and the other mining and mill whistles far and near, up and down the canyon, chipped in to swell the infernal din, aided by the clanger of the fire

Some engine companies sponsored "junior" hose companies, supplying their younger comrades with a jumper and hose. Incorrectly identified as Curry No. 2 of Carson City, this view (circa 1879) shows the juniors of Yellow Jacket Engine Company No. 2, Gold Hill. (Warren Engine Company No. 1)

> bells and the hoarse shouts of the running firemen with their jingling and rattling machines. And the way they did streak it down the long steep declivity of Main Street was a caution to greased lightning. Luckily no one was run over, killed, or had any of their legs or necks broken. The boys with the hose carriage of Liberty Engine Company No. 1, made the cleanest, liveliest and smartest run we ever saw. All hands claimed "first water," but Judge Wright was there himself, and confidently informs us that Lincoln Hose got "first water." He is a member of that company, however, therefore might be considered interested authority. Anyhow, the fire got squelched in a very particularly lively manner after the boys got a fire rake at it.[4]

Once at the fire, the members of a hose company could expect action. First, there was the problem of supplying the engines with a sufficient quantity of hose, then laying hose from the engines to the fire. Quite often it was the members of the hose companies who manned the nozzles and got into the thick of the battle, while others labored on the brakes of the hand engines. Sometimes, there was more of a battle with the hose than with the blaze itself. Eugene Markey had a rather tough bout with a length of pressurized hose at a fire in 1876: "Sometimes the pipe got Markey down and punched and flooded him, and then he would get a collar and elbow hold and flop it over on its back. Then they took turns flopping each other up through the badly damaged roof and both came near going down through the charred and broken floor several times," the newspapers reported. Markey finally got the best of his "live" line, maneuvering it into position and playing a good stream of water on the fire.[5]

Sometimes the hose carriages were the only pieces of equipment in the fire department with the mobility to get to particularly isolated fires, or areas where the terrain was so rough that the engines could not negotiate it. Also, when the hand pumpers were out of service, it was up to the hose companies to fill the gap. In 1871, the hose companies made a particularly fast dash to a fire in Virginia City, because both Washoe Engine No. 4 and Young America Engine No. 2 were out of service for painting and repairs.[6]

Many of the hose companies, though attached to a particular engine company, had their own houses. Invincible Hose Company No. 4 had a small hall near that of Washoe Engine Company No. 4. It was fitted up not only to shelter their hose carriage, but also as a reading room and place for meetings. The members of the hose company were very proud of their reading room, which they stocked with the latest daily newspapers, periodicals, and other reading matter.

The houses of the various hose companies were just as ornamental as those of the engine companies, and could rival them in comforts. Neptune Hose Company No. 5 featured a special roll call of the company made by Mary Banaham and Maggie Sullivan, who also presented the firemen with a fancy meeting flag. Along with helmets, flags, belts, and other decorations, one corner of the hall held a fine service trumpet, which had been made for the company by Philip Flick, a Neptune member who was employed by Gillig, Mott & Company, one of the city's leading hardware establishments. The trumpet was made of brass and copper, with a shield of Comstock silver on one side, bearing an inscription from its maker to the company.[7]

The hose companies were as proud of their service to the city as the engine companies, and they took the criticism leveled at the fire department after the Great Fire of 1875 with the same indignation. As the companies considered disbanding, a note of displeasure from one hose company was sounded in the columns of the *Territorial Enterprise*:

> At a meeting held last night Neptune Hose Company, formerly attached to Knickerbocker Engine Company No. 5, disbanded. They quit on the square—they say they don't owe a cent to any man and don't know that any man owes them a cent. It is a square quit, and they don't know that they will ever again make their appearance before the public.

Shortly thereafter, the company's carriage and equipment were sold and the members formally disbanded.[8]

Unlike the Virginia Fire Department, which had started with the fire engine of Virginia Engine Company No. 1 and then formed hose companies, the Gold Hill Fire Department had its origins in a two-wheel hose carriage. It arrived in 1864, for Silver Bar Hook and Ladder Company No. 1, which changed its name to Liberty Hose Company No. 1 a few months later, when Nevada attained statehood.

The earlier company formed a committee to solicit public subscriptions for the purchase of the cart. A benefit performance of Wilson and Zoyara's Circus was scheduled to raise funds. The press strongly supported purchase of the hose cart and development of fire protection for the town, and the *Gold Hill News* urged its readers to attend the benefit: "The object of the benefit is not merely a laudable one, but one that appeals directly to the sentiment of self-preservation. Let our citizens, one and all, purchase tickets whether their inclinations or business arrangements lead them to attend the circus or not." A few days later the newspaper reported on the success of the event, with a growing fire fund for the purchase of the needed hose cart. Much later, the newspaper bragged that the city was preparing to meet any disaster, with the installation of a new iron pipe system to provide the town with large supplies of water: "With the present prospects of an ample protection against fire, we feel confident that our people already feel easier in their sleep—and 'fire insurance' companies may go to—grass!"[9]

The new hose carriage arrived late in October, amid praise from the press. The carriage, built by Folsom & Hiller in San Francisco, was described by the paper as being

> . . . of a magnificent make, and cost, delivered, $400, including freight. The wheels are five feet eight inches in diameter, with an

> iron framework for the reel, and the carriage is surmounted by a gilt eagle, holding in its beak the silver badge "No. 1." This carriage is capable of holding 1,000 feet of hose on its reel, and the bully company of fifty members, with those first rate firemen, W. H. H. Lee, Jo. Flower, and W. E. Hale as Foreman, and First and Second Assistant Foremen, will prove an effectual safeguard against fire.[10]

If the town was thrilled by the arrival of the new Liberty Hose cart, there was considerable skepticism over the arrival of the hose carriage for Yellow Jacket Hose Company No. 2, in 1865. Like No. 1, the Yellow Jacket Hose carriage was supported by public subscription and a benefit performance at the theatre in Gold Hill. When the necessary funds had been gathered, the new fire company sent a committee to San Francisco to purchase the hose carriage formerly belonging to Washington Hose Company No. 1. A short time after the committee arrived, member G. A. Hart wrote that purchase of the carriage, along with a quantity of hose and other items needed by the company to fight fire, had been completed. He acknowledged the assistance of the president of the Fireman's Trust Bank, the editor of the *Spirit of the Times*, and San Francisco Fire Chief David Scannel in acquiring the carriage, which was to become Gold Hill's second piece of fire apparatus.[11]

Despite the confident air of optimism of the purchasing commttee, critical comment from the *Gold Hill News* struck like a cannonball upon arrival of the new hose carriage:

> The "What is it?"—It came to town yesterday by Fast Freight, from over the mountains. It came from San Francisco. "Washington Hose Company No. 1" sent; and of all the d______d "billiks" that ever came to Washoe, the "What is it" is the biggest! We do not wish to be severe in our remarks—but the people of Gold Hill are not exactly the people to be humbugged.

The carriage had arrived in an abominable condition—broken, bent, muddy, and generally a wreck. After a week of cleaning, burnishing, and painting by the firemen of Yellow Jacket Hose Company No. 2, however, the *Gold Hill News* changed its tune: "We don't believe it is as big a 'billik' as was first reported—and as, in fact, looked to be—and for the honor and credit of the purchasing committee we make this correction."[12]

The hose carriage with the longest and probably the most colorful career in the fire service of the Comstock arrived in Virginia City in

1870. It was put into service by Good Will Hose Company No. 2, attached to Young America Engine Company No. 2. The carriage was purchased in Philadelphia for $460, and F. A. Tritle arranged for free shipping over the transcontinental railroad. The new hose carriage was expected to arrive within three weeks of its purchase, but it did not appear. On September 7, 1870, the newspapers announced that the carriage could be looked for "within a day or two"; but on October 26, they reported that the new hose carriage had mysteriously disappeared. They speculated that Virginia City, Montana, might accidentally have received Good Will's shipment.[13]

Had the hose carriage been kidnapped? Telegrams were sent in all directions in an effort to discover its fate. On November 1, the *Gold Hill News* reported the new carriage had been found hiding in Truckee, California, where it had been mistakenly delivered to surprised firemen, who gladly accepted the "gift" from an unknown benefactor. A few telegrams and letters later, the mixup was settled, the "muchly miscarried hose carriage" was reported on its way to the Comstock. On November 14, 1870, the new carriage arrived in Virginia City. The *Territorial Enterprise* heralded the arrival with a brief history of the cart, which had served the Good Will Hose Company of the Philadelphia Fire Department, and had been "jerked about remarkably lively at both fires and fights." Indeed, the delicate looking carriage had even been thrown into the Schuykill River by a rival fire company while members of the two companies battled each other during a fire. The carriage was capable of holding eight hundred feet of hose, and was painted blue with gold leaf detail. It featured two fine brass fire bells and two lamps, with the side badges of the cart bearing the name "Good Will." On the round toolbox in the front of the cart was a painting of a dog seated in front of a burning building.[14]

The ornate carriage, with high, fancy ironwork, was housed in the hall of Young America Engine Company No. 2 on C Street. The members of Good Will Hose Company dashed it along the streets at the first alarm of fire, and took it to drills. In 1875, the carriage was called out for duty during the Great Fire that leveled most of the town. After the fire, it was cleaned and rehoused in No. 2's house, which had escaped destruction, but its days of service in the Virginia Fire Department were numbered. With formation of the new paid fire department, Young America Engine Company No. 2 disbanded and reorganized as a building and real estate association, selling all of its equipment. The hose carriage was sold to

The oldest known piece of fire-fighting apparatus in Nevada is this four-wheel hose carriage built in 1839 for Good Will Hose Company No. 25 of the Philadelphia Fire Department. It was purchased in 1870 by Good Will Hose Company No. 2, attached to Young America Engine Company No. 2, which sold it to Liberty Engine Company No. 1 after the Great Fire of 1875. The carriage was in service until 1938; it has now been restored and is on display in the Comstock Firemen's Museum. (Photo taken in 1938 by Walt Mulcahy)

Liberty Engine Company No. 1 of Gold Hill, to replace the old two-wheel hose cart in use since 1864.*

The Butt Enders promptly put the hose carriage back into service, and organized a new group of men to run with it to fires. The carriage was repainted red to cover scorch marks and nicks acquired during years of hard use, which entailed bouncing along the steep, rocky streets of Virginia City, and occasionally slamming against a building or other object. Brightly trimmed in blue, white, and gold, the carriage saw many fires in Gold Hill, receiving the same kind of rough treatment on the run to the fire. It was even dragged across the high Crown Point railroad trestle to a fire near Baltic Switch.[15]

After the hand engine of Liberty Engine Company was sold in 1910, Liberty's hose carriage became the primary fire-fighting apparatus of Upper Gold Hill, assisted at times by the hose carriages of Divide Hose Company No. 2 and Lincoln Hose Company No. 3 in Lower Gold Hill. Although there were no more musters or firemen's tournaments, the carriage was used in at least one contest in 1881, against the hose carriage of Yellow Jacket Engine Company No. 2, for the prize of a silver serving set. Liberty won the match and a rematch for a silver trumpet.[16]

When Liberty Engine Company No. 1 formally disbanded in 1938, the hose carriage and the little jumper of Junior Hose Company No. 1 were the only apparatus left in the engine house. They remained stored until 1952, when heavy snows and the weight of school desks stored in the attic of the engine house caused its collapse. The carriage was badly damaged, its ironwork crushed, hose reel smashed, and at least one wheel crippled. The jumper was taken to the courtyard of the Storey County Courthouse until it was restored in 1878 and placed in the Comstock Firemen's Museum in 1979. The Liberty Hose carriage, however, was stored until 1976, when it was repaired and painted. It won the 1976 Nevada Day Parade Sade Grant Memorial Award trophy. In 1978, while returning from another Nevada Day parade, in which it won first-

*When the Virginia Fire Department disbanded, virtually all of the volunteer companies liquidated their assets and split the funds derived from the sale of fire apparatus and property among the members. Young America Engine Company No. 2 was comprised of several major Comstock businessmen who had directed the company's funds in such a manner that the company owned two or three rental properties in town which provided additional funds for the company. After disbanding, the company formed a building and real estate association in which funds from the sale of fire apparatus were re-invested in property and businesses.

place historical entry honors, the carriage fell off a flatbed trailer and was dragged for nearly one-half mile along the highway south of Silver City. The damage was almost identical to that it suffered in 1952.

Using photographs of the carriage taken during its days of service in Gold Hill, the volunteer firemen of Virginia City once again restored the ornate hose wagon. Original parts removed during the 1976 restoration revealed a coat of blue paint underneath the red, confirming the history of the carriage. In 1979, it was housed with other hose carriages in the Comstock Firemen's Museum, and in 1980 was accorded "Best of Parade" honors at the West Coast Championship Firemen's Muster in San Jose, California. At that gathering of antique fire equipment, Comstock firemen ran the old Liberty Hose carriage and pulled hose from it in a demonstration that marked the first time the cart had run with hose and firemen since its retirement in 1938.

Another hose carriage with a long and colorful history was that of Divide Hose Company No. 2, now housed along with No. 1's carriage in the Comstock Firemen's Museum in Virginia City. The four-wheel carriage was originally purchased from fire engine manufacturers Button & Blake in 1879, to replace the old hose carriage of Yellow Jacket Engine Company No. 2, which had been purchased in San Francisco many years earlier. The carriage remained in service with the Jackets until they disbanded in 1885.[17] During this time, it challenged No. 1's carriage in a special hose cart race, with a silver serving set for a prize and a rematch prize of a fireman's trumpet. The heavy carriage of No. 2 was no match for the Butt Enders, however.

When Gold Hill Fire Chief J. F. Gladding formed Divide Hose Company No. 2, on the Divide between Gold Hill and Virginia City in 1885, the old Yellow Jacket carriage was taken from storage and placed in the care of the new company. Originally painted white with fancy gold leaf trim, it was repainted deep red by the new company, with gold leaf trim and powder blue shading. The first two years of the new hose company were filled with controversy, as disagreements flared over the company's response to fires in Virginia City. Fueling the argument was a petition from Divide Hose Company seeking permission to move its quarters from Gold Hill to a new house for the carriage which was actually over the boundary line in Virginia City. The motion was denied by a three-to-four vote of the delegates of the Gold Hill Fire Department. It was the determination of the board that it "could not allow a hose or engine company who has their building in Virginia City to receive their donation from the funds of the Town of Gold Hill as a company

The hose carriage of Divide Hose Company No. 2, shown in the Divide firehouse in 1937, was built in 1879 by Button & Blake for Yellow Jacket Engine Company No. 2. When the Jackets disbanded in 1883, the carriage was put into storage. In 1885 it was turned over to the new Divide company. It fought its last fire in 1942. (Author's collection)

belonging to the Town of Gold Hill." Within a month, however, promises were made, compromises agreed upon, and a new vote on the matter sought: "After a number of remarks made by all members present, it was moved, seconded and carried that the Board of Delegates do approve and acquiesce of the removal of the Divide Hose Company No. 2 to their new hose house over the Virginia City line, still belonging to and being of the Gold Hill Fire Department."[18]

Divide Hose Company No. 2 responded to fires in both Gold Hill and Virginia City, working with the Virginia Paid Fire Department in trying to keep the two towns from repeating their fiery pasts, and receiving praise in the local newspapers for their efforts. The company even responded, along with Liberty Engine Company No. 1, when the International Hotel burned on a cold winter's night in 1914, and made the run all the way from the Divide on icy, slippery streets.[19]

In 1937, the surviving members of Divide Hose Company No. 2 voted to disband. After taking a photograph of the carriage in the hall, they closed and locked the doors. The firehouse, where the Divide volunteers had often gathered for meetings and lively discussions, and which had one of the few telephones on the Divide, used frequently by the citizens to call doctors and relatives, was virtually abandoned. But it was not to be the last fire call for the old hose carriage.

In 1942, a fire started by youths in a car on Ophir Grade swept over the Divide. Firemen and citizens scurried to fight the spread of the flames, but they were fanned by a strong wind and spread rapidly. The fire department had every length of hose in service that night, but more hose was needed. Someone recalled that the Divide Hose carriage, stored nearby in the old firehouse, still had several hundred feet of old hose on it, so it was called into service one last time. The carriage was taken out of the firehouse to a lot across the street, where the hose was rolled off. Shortly afterward, the windswept flames levelled the old home of Divide Hose Company No. 2, leaving only the carriage and a very few belts and helmets as reminders of the pioneer fire company.[20] According to those living on the Divide at the time, the people responsible for the fire had not been invited to a party being held that night, and some believed the fire was set out of spite. The young men were never held accountable, despite the destruction of several automobiles, ten homes, and about thirty other buildings.[21]

With the formation of the Virginia Paid Fire Department, after the Great Fire of 1875, came the first horse-drawn fire apparatus on the Comstock Lode. Identical hose carriages were purchased from the

The City Hose Sleigh, built in 1877, was a unique piece of fire apparatus for Nevada. Few towns had enough snowfall to warrant such equipment. The sleigh was sold to the Pierce Miller collection in Modesto, California, and is still displayed there. (Author's photo)

Kimball Manufacturing Company in San Francisco. City Hose Carts No. 1 and No. 2 served the community for more than forty-five years, together with a horse-drawn sleigh-type carriage used during winter months when heavy snow made travel by wheeled vehicles impossible. The late William "Bud" Spargo, who reeled hose back onto the City Hose carts after fires as a boy, recalled that snow was packed into the floor of the Corporation Fire House and allowed to freeze over. This provided a slick running surface, so the sleigh could get a good start out of the firehouse when responding to fires.[22]

City Hose Cart No. 2 and the sleigh were acquired in 1933 by Pierce A. Miller, a private carriage collector in Modesto, California, and are housed today in his collection. City Hose Cart No. 1, the first piece of horse-drawn fire apparatus ever used on the Comstock, remained in service until the acquisition of the Comstock's first piece of motorized firefighting equipment in 1932. In 1948, the Storey County Commission presented City Hose Cart No. 1 to the fledgling Nevada State Museum in Carson City, where it was stored outside or in barns for a number of years.

As the centennial of Warren Engine Company No. 1 approached in the early 1960s, the carriage attracted the attention of Bernard Sease, a volunteer who would later become chief of the Carson City Fire

Department. Sease and others endeavored to have the carriage loaned to the Warren Engine Company Museum, and began initial restoration. However, other duties and projects forced a cessation of the restoration effort, and the cart was placed in a State Museum storage facility in Carson City, where it was virtually forgotten until the summer of 1977, when research revealed its existence. Representatives of the fire department in Virginia City requested that the carriage be returned to its original home on the Comstock Lode. The Board of Trustees of the Nevada State Museum agreed to the loan, and City Hose Cart No. 1 was loaded on a trailer for its trip back home.

In 1979, after several months of restoration work guided by Bud Spargo, who worked on the carriage at his Virginia City shop, City Hose Cart No. 1 was housed in the Comstock Firemen's Museum with the other historic hose carriages that have served the Comstock. In that same year, the carriage was pulled by a team of matched white horses in the Nevada Day Parade and won the coveted Governor's Trophy, which was presented to the volunteer firemen by Governor Robert List. It was the first time since its retirement in 1948 that the hose carriage had been hitched to a team of horses. In 1981, the carriage traveled to San Francisco, where it had been manufactured, and participated with the Liberty and Divide hose carriages in special ceremonies commemorating the seventy-fifty anniversary of the 1906 earthquake and fire. In a special demonstration, City Hose Cart No. 1 laid nine hundred feet of hose, supplying hose lines to three steam fire engines as well as laying a hose lead for members of Liberty Engine Company No. 1 to squirt as if they were fighting a fire.

XI. *Fighting Fire With Fire*

It is hard to imagine the firemen of Virginia City and Gold Hill hauling heavy steam fire engines up and down the steep streets of the Comstock by hand. The distance to fires was so short that the various fire companies deemed it impractical to maintain stables of horses to haul the heavy machines to fires.

Nevada received its first steam fire engine in 1872, with the arrival of a shiny new Silsby steamer for Young America Engine Company No. 2. The steamer was a thing of wonder for the firemen and citizens of the Comstock, who were used to seeing forty or fifty men working the brakes of a hand engine at a fire. A new era had arrived in fire fighting on the Comstock, and the members of Young America were proud of their new machine and its promise of good service to the community.[1]

The new steamer cost $5,000, with a considerable portion of the funds coming from the treasuries of Virginia City and Storey County. The towns stipulated that Young America Engine Company take the engine to Gold Hill in the event of a fire there. Once the engine was completely assembled, with all its brass and nickel polished, the members of the company hauled their new prize out onto C Street for a photograph in front of the Presbyterian church across the street from their engine house. Afterward they put the new machine on display for public inspection. More than two thousand persons visited the Young America hall that day to see the steam engine work, but due to a lack of high-pressure, carbolized hose, there was no squirting done for the assembled crowds. The firemen did throw light combustible materials into the firebox in order to show the machinery in action, which delighted the many spectators assembled in front of the engine house.[2]

Nevada's first steam fire engine was an 1872 Silsby rotary steam engine purchased by Young America Engine Company No. 2. Members of Young America posed with their engine in front of Virginia City's Presbyterian Church in 1872. (Nevada Historical Society)

The Silsby steam fire engine, which had been manufactured in Seneca Falls, New York, weighed 4,500 pounds when fitted out and ready to fight fire. It was decorated with fine lamps, which had cut glass panels bearing the name of the company. An ornate brass lamp adorned the top of the shiny air pressure dome, with colored glass bearing the number of the company. Painted in bright letters with brass cutouts on the front of the frame beneath the machinery was the name "Young America."[3]

The shiny brass and moving parts failed to convince some of the old skeptics of the fire department. After an incident at a fire on the Divide, shortly after the arrival of the "infernal machine," they were convinced that the new steamer was inefficient. There was a lack of high-pressure hose in the fire department, and the old leather hose could not withstand the high pressure produced by the steamer, so Young America had maintained a very low pressure at the fire for fear of bursting the old hose.

While the skeptics howled, and the members of Young America Engine Company anxiously awaited new high-pressure, carbolized hose from San Francisco, the newspapers were supportive of the fire-fighting innovation. They pointed out that although any hand engine in the fire department could have matched the distance thrown by the steamer, it would have been hard pressed to maintain that stream as long as the steamer, which needed only steam and water to throw water, instead of men to work brakes. The newspapers promised an improvement in the engine's performance once the appropriate hose arrived in town.[4]

The next performance was indeed an improvement over the 180 feet thrown at the Divide fire. No. 2 hauled its new steamer to Gold Hill, where it was fired up and squirted a steady stream over the high Crown Point trestle of the Virginia & Truckee Railroad. After that, they moved the engine to the large cistern on Bowers Grade and threw two solid streams over great distance for an extended time. Then they moved to the cistern in front of the Liberty Engine Company hall on Main Street and sucked it dry. The streams thrown by Young America's steamer, the *Gold Hill News* said, "were better than we ever saw thrown by hand engines, and the beauty of it was that they were kept up for any length of time required—as long as water and wood could be found." For the men of Young America Engine Company, however, that trial was not enough to convince skeptics of the town. Two days later they hauled the steamer out for another trial, to settle once and for all the question of the efficiency of the new engine. As the *Territorial Enterprise* reported:

The steamer of Washoe Engine Company No. 4 was a Button similar to this.

> She threw a single stream of water 40 feet above the spire of the Catholic church. The perpendicular distance from the ground to the top of the spire is 179 feet. At a second trial the steamer threw two streams 10 feet above the spire, through 1½-inch nozzles. After returning to South C Street the engine was subjected to another trial, during which a stream was thrown from C to B street [uphill] over the roof of Mr. Fair's house. Great satisfaction was expressed by all who witnessed the splendid work of the steamer.[5]

It took about seven minutes to get up a good head of steam in the boiler, even using light combustible fuels. The innovative members of Young America Engine Company No. 2 had steam piped into their engine house and connected to their steamer, so that a warm boiler was always ready in the event of a fire alarm. The skeptics had been quieted; the steam fire engine was on the Comstock to stay. During the month following the tests of No. 2's steamer, the companies in the Virginia Fire Department and the Gold Hill Fire Department were racing furiously to buy steam fire engines. Even Carson City's firemen began efforts to acquire a steamer for a new company which was to be named after organizer and Ormsby County Sheriff Shubal T. Swift.[6] That engine is today displayed as part of the Harrah's Automobile Collection Pony Express Museum, in Reno.

On the Comstock, Washoe Engine Company No. 4 and Knickerbocker Engine Company No. 5 were the next to order steam fire engines. No. 4's engine arrived in February, 1873, by way of the Central Pacific Railroad. When the engine rolled into Carson City aboard the Virginia & Truckee Railroad, railroad superintendent Henry P. M. Yerington telegraphed the news to Virginia City. The new steamer, still aboard a Central Pacific flatcar, rolled boxed and covered through Gold Hill on its way to Virginia City, revealing only a portion of the wheels. It was to remain crated and partially dismantled until the arrival of the manufacturer, Lysander Button, from Waterford, New York.[7]

After Button finished assembling the engine, the Washoe firemen scheduled a grand trial on February 17, 1873, a Sunday. They were disappointed when Button told them he would bring the engine out on the Sabbath only in case of fire, he being from a part of the country where the day was more strictly observed than in the rough and often wild mining towns of Nevada. The *Gold Hill News* observed of the situation, "One of our Gold Hill saloon keepers hearing of this, remarked that it was all very well, but an Eastern man could not remain long in Virginia or Gold Hill without learning to drink, smoke, play cards and run foot races on Sunday." The delayed test, however, proved most satisfactory to Washoe Engine Company No. 4. The new steamer threw a stream 252 feet, through a 1⅛-inch nozzle, topping the best stream of a hand engine on the Comstock, which could throw only about 209 feet.[8]

No. 4's steamer was as fancy and efficient as that of Young America Engine Company No. 2. It cost $4,000, and weighed 5,000 pounds when outfitted and ready to battle a blaze. It was a double-piston fire engine with two discharges and a capacity of 372.322 cubic inches, or 1.61 gallons to a revolution. The engine was balanced in such a manner that there was no shaking of the running gear while at work, and hung on half-elliptical steel springs. To move it along the rough streets of Virginia City, there were 54-inch diameter steel wheels in the front, and similar wheels, 66-inches in diameter, in the rear.[9]

Knickerbocker Engine Company's new Amoskeag steamer received a real trial before it ever arrived on the Comstock, on January 7, 1873. It was turned out of the Amoskeag shops in Manchester, New Hampshire, just in time to be called into service at Boston's great fire of 1872, for

The Amoskeag steamer of Knickerbocker Engine Company No. 5 got a real "baptism of fire" when fighting the Great Boston Fire of 1872. The engine arrived in Virginia City still grimy from the Boston fire but was soon in service. After the company disbanded in 1877, the engine was sold to Virginia City's new paid fire department, which rarely used it and sold it to Gilroy, California. It was put into service there, as seen in this 1917 photo. (Gilroy Historical Society)

what the Comstock's newspapers called a real "baptism of fire." The new engine, which cost nearly $5,000, including freight charges, was not shiny and bright in appearance when it arrived in Virginia City, due to its services in the Boston fire. But within a week, the ambitious firemen of No. 5 had its brass and nickel gleaming.[10]

No. 5's new steamer got its first Comstock trial one week after its arrival. The event was attended by a great number of the townspeople, including *Gold Hill News* editor Alf Doten, who reported the engine threw a stream 241 feet through 1,000 feet of hose, using a 1¼-inch nozzle. Afterward, the engine took water from a cistern on north B Street, throwing a stream over the houses from B to the middle of E Street. Doten wrote, "With a little more power they might station their machine in the middle of the city and drown out any ordinary fire which might occur in any portion of the city without running to it." While normal steam-up time was seven minutes, No. 5 always kept the water in its boiler warm and cut steam-up time to only four minutes. Once steamed up, the 6,310-pound Amoskeag could produce two solid streams of water.[11]

The Divide, between Gold Hill and Virginia City, was next to receive a steam fire engine, with the purchase of a Clapp & Jones steamer from Hudson, New York, for Monumental Engine Company No. 6.[12] Yellow Jacket Engine Company No. 2, of the Gold Hill Fire Department, was the last of the Comstock fire companies to purchase a steamer, a Jeffers, which arrived a short time after No. 6's. Shortly after the Jackets received their machine in 1874, a grand trial of the two engines was scheduled in Gold Hill. Monumental Engine Company rolled its giant steamer to No. 2's house, where the two companies joined in some pre-trial refreshments. Then the two engines were taken to the Rhode Island Mill, where there was plenty of water and room for the drill. Monumental fired up first, taking only 7½ minutes to gain 20 pounds of steam pressure. Taking water at a tank near the mill, No. 6 threw a stream about 200 feet up the street through a 1¼-inch nozzle, following that test with a demonstration through a 1-inch nozzle.

Yellow Jacket No. 2 had 20 pounds of pressure in only 7 minutes and, using the same sized nozzle, threw nearly as far as its antagonist. Some witnesses claimed that it threw fully as far, but due to a strong wind, there was no way of accurately measuring the distance thrown by either engine. However, the Monumental steamer was a size larger, being a second-class Clapp & Jones weighing 6,300 pounds without water or fuel. Loaded and ready for action, the Yellow Jacket engine weighed only 5,570 pounds and was a third-class Jeffers steamer built in Pawtucket, Rhode Island.

Following the drill, the companies adjourned to the residence of John Jones, who provided refreshments for the firemen. Then Monumental members hitched their hand-drawn steamer to a team of horses provided by Thomas Gallagher, and made the trek uphill to their engine house on the Divide. They took with them a large delegation from Yellow Jacket Engine Company No. 2, who were treated to yet another party.[13]

As exciting as the trials of the steamers was the housing of the new engines. When Yellow Jacket Engine Company received its new steam fire engine, a grand affair was organized which included plenty of beer, and a long table spread with boiled ham, cheese, bread, and other food.

Chief Abe Jones of the Gold Hill Fire Department called the meeting to order and made a welcoming speech to those gathered to inaugurate the new steamer. John McDonnell, president of the Gold Hill Miners' Union, addressed the crowd, then recited the poem, "Sheridan's Ride." Virginia City Fire Chief J. P. Bell also spoke, and ended his address by singing the Celtic ballad "Tim Flaherty." He was followed in song by W. Brennan, singing "Castles in the Air." Parson Gibson rounded out the ceremonies with several "good old camp meeting songs" and a brief prayer. After another onslaught on the refreshment table, the crowd witnessed the first firing of the new steamer, watching it work inside the engine house of No. 2.[14]

The age of the steam fire engine on the Comstock really lasted only as long as the old volunteer fire department. When the Virginia City volunteers disbanded after the Great Fire of 1875, and the Virginia Paid Fire Department was organized, the exodus of the steamers began. One by one they were sold—Washoe Engine Company No. 4 sold its Button & Blake, Knickerbocker Engine Company No. 5 sold its Amoskeag to the city, and the state's first steam fire engine was sold by Young America Engine Company No. 2. The former steamer of No. 5, though seldom used, remained on duty with the new paid fire department through the late 1880s, but was eventually sold to the Gilroy, California, Fire Department. Only Monumental Engine Company No. 6, on the Divide, kept its steamer in service for response to either Virginia City or Gold Hill fires. Yellow Jacket Engine Company No. 2 responded to fires in Gold Hill until that company disbanded in 1882. Its steamer was stored in the Corporation Fire House for a number of years before it was sold. Virginia City's town fathers issued the final blow to the steamer by charging the Gold Hill Board of Fire Delegates rent for its storage at the paid fire department.[15]

After years of service, the consolidated county government decided that it was inefficient to maintain Monumental Engine Company No. 6, which had been receiving county and city subsidies. County officials argued that with the new water system providing a virtually inexhaustible supply of water at tremendous pressure there was no need for a steam fire engine, especially one run with the added expense of a volunteer fire company attached. Monumental Engine Company No. 6, realizing the city and county would no longer contribute to continue its service, disbanded and offered its equipment for sale. In 1887, the company sold its Clapp & Jones steamer, with all of the company property, including uniforms, helmets, belts, and other equipment, to

These firemen in red shirts and leather helmets fired the Jeffers steam fire engine of Yellow Jacket Engine Company No. 2 in Gold Hill. It took a lot of strength to move the heavy steamers around the hilly streets of Virginia City and Gold Hill, and then to pump water until the supply ran out. (Nevada Historical Society)

Monumental Engine Company's Clapp & Jones steamer after it became a horse-drawn vehicle and after its arrival in Reno in 1887. The engine remained in service in Reno until 1903, when it finally wore out and was scrapped. (Comstock Firemen's Museum)

Reno's Washoe Engine Company No. 2, which converted the steamer to allow it to be hauled by horses. The steamer remained in service until 1903, when it finally wore out. The *Nevada State Journal* lamented the passing of "Old No. 6," with the following story:

> Take her away for $150. Wrench off the jacket, break up the boiler with sledges for recasting, try to fit the wheels to a baggage truck. Take her away. She's yours for $150.
>
> It's all very well to dispose of old No. 6. She has passed her era of usefulness. There is a bulge to the boiler and the axles are out of line. She is too long coupled to make a short turn and she is slow getting up steam.
>
> Old No. 6 belongs to the past generation. She is an old fogy. There is no place for her in the present days when grey heads are weeded out and the race is not to the rheumatic limbs.
>
> What of old No. 6's history? What of the men who ran to fires proudly conscious that the steamer was thundering up the street behind them and would in a moment be conquering the fire that was even then bursting through the roof and windows and coloring the night with lurid tinge?
>
> The men who ran with No. 6 in those golden days are now in the churchyard or faltering toward that land of rest.
>
> No. 6 was once the pride of the Comstock and of the state. When that magnificent piece of machinery was taken up to the old Lode she was resplendent in nickel and brass and was the handsomest fire engine that had yet been turned from the shops. She cost a cool ten thousand dollars, but was well worth it.
>
> Many a fire on the Comstock was stopped in its incipiency by No. 6 and many a night she worked to the exhaustion point to confine a conflagration to the limits in which it originated.
>
> No. 6 was a willing worker. She never complained. She was not a cause of expense, even after she was purchased by the city of Reno and installed in the new engine house here.
>
> But like the Deacon's One Hoss Slay, there was an end to it all and a sudden end. She collapsed a few weeks ago when struggling to the limit of her powers to conquer the Sierra Street fire.
>
> Who wants old No. 6? Who will give $150 for her? To the boneyard with the worthless hulk![16]

XII. A Deed Most Foul

"There is a bit of stout rope with a noose at the end for the scoundrel if he is ever caught at it," was the warning the *Gold Hill News* issued in 1870, after a wave of fires started by an arsonist.

From the early days of the Comstock, arson was a major problem for the sprawling mining communities of Gold Hill and Virginia City. For several years reports of incendiary fires were common, so frequent as almost to be expected. There were various methods employed in these incendiary efforts; splashing coal oil on dry buildings, or setting wood piles, trash, hay, or other materials on fire with a match. The problem was so serious that, in Virginia City alone, there were at least five arson-related fires each year, with a scattering of similar incidents in the neighboring towns of Gold Hill and Silver City.

A small sampling of the many cases of arson fires on the Comstock, over a period of about six years, presents a good idea of the seriousness of the problem:

November 19, 1863 – Judge Wetherill convicts and sentences one of the few Comstock arsonists to be caught. J. C. Bodman had his examination before the judge and was held to answer to the next grand jury in the sum of four thousand dollars. In default of that sum, he was committed to jail. Bodman had set fire to the Summit Lake House, which was totally destroyed, along with several smaller adjacent buildings.

December 4, 1865 – The southern gable of the Vesey House in Gold Hill was set on fire by an arsonist, but was extinguished with a few buckets of water. The *Gold Hill News* warned the incendiary to "vamoose the ranch, or he may get a rope around his neck."[1]

November 12, 1868 – "Coal oil incendiarism" was blamed for the destruction of the St. Louis Brewery on C Street. The firemen battled through the night, but despite their efforts, the interior of the building and its contents were entirely destroyed, leaving only the walls standing. A strong wind and chilling rain hampered the efforts of the firemen, but the blaze was confined to the brewery where it started. It was not the first time attempts had been made to burn the building, and firemen inside the burning structure noted the strong smell of "rank coal oil suffocating and almost strangling the brave firemen and others with whom it came in contact." The *Gold Hill News* suggested, "Whatever was the motive inducing this infernal crime, it is to be hoped that the perpetrator may be ferreted out and brought to the halter he so richly merits."[2]

January 5, 1869 – The saddlery and harness shop of J. P. Smith, in a small stone building at C Street and Sutton Avenue, was destroyed between three and four a.m. The interior and the contents were totally destroyed. The basement of the building was occupied by Chinese, who reported hearing somebody carefully walking overhead about three a.m., shortly before the alarm of fire.[3]

May 4, 1869 – "An attempt was made to fire the large wooden building on South C Street, Virginia, known as the 'Upper Level,' " the *Gold Hill News* told its readers. The fire was discovered by a woman who ran out and extinguished it before it had a chance to spread. The cause of the fire was the piling up of "a lot of combustibles saturated with coal oil against the rear end of the house." Again, the newspaper issued a stern warning: "If that fellow tries the thing on again, he will get shot sure."[4]

August 23, 1869 – The wooden building formerly housing Darwin's Assay Office on Taylor Street, between C and D streets, was set on fire. For a time it threatened to destroy every building around it. Not only did the lofty balcony of the Black and Brothers building opposite catch fire from the extreme heat several times, but several adjacent buildings were filled with smoke and thought at one time to be on fire. "The fire is fully believed to be the work of an incendiary," the newspapers said the next day.[5]

August 28, 1869 – Several buildings at the intersection of F and Taylor streets were burned and scorched by a fire "set by some mischief making scoundrel." Not only was the south side of St. Paul's Episcopal Church badly scorched, but the bell deck was completely destroyed, and the pews, furniture, and other decorations were badly damaged. The small dwelling adjoining the fire, occupied by W. T. O'Neil, was totally destroyed, along with the building in which the fire originated. Fortunately, however, quick-acting firemen removed the contents of the

O'Neil dwelling before it was consumed. Dr. Hiller's residence nearby was also badly scorched by the fire.[6]

July 6, 1870—"Some villain tried to set Piper's Opera House on fire about two o'clock this morning," the *Gold Hill News* reported. "A lot of camphene, and other combustible material was distributed beneath the northwest corner of the building by crawling under the sidewalk, and when discovered the flames were already bursting forth, the privy of the station house, which is in the basement of the theater, being on fire." Quick action ended the fire before the hand pumpers could be brought into work. The newspaper issued yet another warning that "Somebody must stop their fooling, or take the consequences."[7]

November 25, 1870—The *Gold Hill News* stated:

> About 1½ o'clock night before last, and again at 5½ o'clock Saturday morning some rascal set fire to the old Mexican hoisting works at Virginia, calling out the whole Fire Department each time to stop the conflagration. That scamp will be caught at his tricks yet and properly lynched. This is his third attempt on that building.[8]

In February of 1871, however, began a series of events that stretched over several months and finally brought the arsonist to justice. During the week of February 13, several fires were reported in the *Territorial Enterprise* and the *Gold Hill News*. According to the newspapers, the fires suggested an organized and persistent determination to entirely destroy Virginia City by fire. Only two weeks earlier, a major portion of the city had been destroyed as a result of the arsonist's activities, leaving thousands of dollars in damage and two persons dead.[9]

On February 12, at about three a.m., a fire was discovered just beneath the sidewalk and front of a carpenter shop north of the Collins House on B Street. The fire had been started with a lot of wood chips, shavings, boards, and other materials, saturated with coal oil. At about the same time, a similar fire was discovered and extinguished in a wooden building near Milatovich's store on south C Street. The arsonist had shoved a bunch of paper into an aperture at the head of an old stairway and started it blazing when it was discovered. Only the quick efforts of the fire department and those discovering the blazes saved the town from a repetition of the disaster which had swept part of the business district away only two weeks before.[10] Again, the *Gold Hill News* issued a stern warning:

> Whatever may be the motive of these black-hearted villains, let them bear in mind that some time or other they may get caught,

> when a short, swift and terrible vengeance awaits them. The police are on the alert, and so also is every good citizen. There are many fellows loafing and lurking about the city with no apparent occupation who will bear watching.[11]

On March 13, someone attempted to set fire to Piper's Opera House, breaking into the building through a back window and saturating some old lumber and firewood with coal oil. Fortunately, passersby saw the flames and were able to extinguish the fire with a few buckets of water before it could spread and do much damage. This time, however, there was a suspect in the attempted arson. Within an hour, Chief of Police George Downey, who was also a member of Nevada Hook and Ladder Company No. 1, had William Willis in jail under suspicion of setting the fire. Coal oil was found on his hands and on the sleeves of his coat, and there were traces of dry lime on his pants, apparently from sitting on a pile of lime on a low platform near where the fire had been started.[12]

The evening before, Willis had attempted to "spar" his way into the opera house, but managed to elude officers Potter and William Stout. Piper. Willis left, but not before publicly swearing vengeance against Piper. On the night of the fire, he was seen lurking about the back of the opera house, but managed to elude officer Potter and William Stout. Shortly afterward the blaze was discovered, and after the two officers consulted with Chief Downey, Willis was captured. The young carpenter, who was a member of Invincible Hose Company No. 4, attached to Washoe Engine Company No. 4, was placed in a cell in the city jail along with Arthur Perkins, also a member of Invincible Hose Company. Perkins had been jailed earlier in the week on a charge of murder.[13]

The next night, as promised so often in the newspapers, vigilantes, locally known as the "601," stormed the jail and took young Willis with them. The "601" were armed with rifles and pistols, with their faces hidden behind masks of red or white cloth. They were gone with Willis for little more than one-half hour, then returned their prisoner and reincarcerated him in his cell, leaving him handcuffed, with the keys in the cell lock.

Willis told his jailers that he had been taken to Piper's Opera House, where many other vigilantes had gathered in the basement. A rope was put around his neck and the other end passed over a beam, within sight of the charred embers from his incendiary attempt of the night before. Willis was forced to confess to that crime, as well as to other arson fires for which he was responsible. In so doing, he revealed to the vigilantes that there were others involved in an arson ring in the town, men who were members of the fire department. Willis told the vigilantes that earlier in the day he had gained access to the opera house by pretending that someone had asked him to deliver a piece of stage equipment. Passing beneath the stage to the basement, he opened the rear shutters of the window, through which he later reentered the building. He admitted that he had later used coal oil to set fire to the opera house while a performance was in progress. Among other fires Willis recalled were the destruction of the Athletic Hall and surrounding buildings, the hall of Invincible Hose Company No. 4, the fire under the sidewalk next to Flurshutz's confectionery store, and the cabin burned near the Provost Guard Station some months earlier.[14]

Willis identified Charles McWilliams, Thomas Laswell, and Arthur Perkins as members of the arson ring, and named Perkins as the ring leader. He told the vigilantes that Laswell assisted Perkins, and McWilliams acted as watchman, although McWilliams had also set some fires on his own. Laswell and McWilliams were then arrested and placed in cells next to their accomplices.[15]

Two weeks later, the inevitable happened. An army of about eighty men forced open the doors of the jail and demanded the keys to the cells from Sheriff Atkinson and Undersheriff Stoner. When the lawmen refused, the vigilantes began an intensive search of the jail, and finally discovered the keys. They went to Perkins's cell. Perkins told the intruders they were doing wrong, but they ignored him. He was asked about his arson activities, but claimed innocence. As Perkins struggled to put his boots on, his interrogators announced that he wouldn't need them where he was going.

Meanwhile, a masked man with a gun awakened Sam Wyckenheim, steward of Young America Engine Company No. 2, told him to keep anyone from ringing the bell on the engine house, and handed him a revolver to help accomplish his chore. Wyckenheim recognized the man as one of the vigilantes, and the bell was muffled.

The masked men marched Perkins to the Ophir Mine, where there was a single-story frame building with a roof beam that extended out about ten feet. From the end of the beam, over a track for ore cars, hung a thick rope with a short chain attached. *The Gold Hill News* reported:

> On top of this track was laid a board upon which the victim was made to stand, while the Vigilantes, evidently not caring to make use of the dangling rope, threw another rope they had brought with them over the beam, and adjusted the end about Perkins' neck. The

> board on the track was then evidently knocked from under him, and Perkins swung into the trackway, hanging by the neck, and passed into eternity—almost as quick as his poor victim William Smith, recently did with the deadly bullet crushing through his brain.

The body was left swinging in the wind for several hours, until midmorning, as a warning to other would-be arsonists.[16]

Willis entered a plea of guilty to first-degree arson, making a trial unnecessary. On April 5, 1871, Judge Richard Rising sentenced him to twenty-one years in the Nevada State Prison in Carson City. Willis took his sentence cooly, and made a short statement that his confession and implication of Perkins, Laswell, and McWilliams was not made voluntarily. "He did not say they were not true, but intimated that they were given under the influence of force or fear," the reporter for the *Gold Hill News* wrote. Willis was sent to prison, and a short time later was captured while trying to escape. Additional time was added to his already long sentence, which he served without further incident. Upon his release, he disappeared. Others arrested as a result of his confession and an investigation were fined and allowed to go free, provided they would leave town swiftly and never return.[17]

Gold Hill News editor Alf Doten, also a member of the shamed Washoe Engine Company No. 4, wrote, "There are black sheep to be found in all companies, therefore it is wrong to condemn this excellent and most serviceable company indiscriminately." Willis, Perkins, McWilliams, and the others, had been drummed out of the company in disgrace, Doten told his readers, and by this act the fire company had been "purified." Invincible Hose Company No. 4 was disbanded by the engine company. Doten wrote, "When properly regenerated Washoe Engine Company will be as good as ever. Their machine is one of the best in the department—never out of service, but always prompt on hand, actively handled and well-mannered."[18]

The members of Washoe Engine Company No. 4 had been wounded by the scandal, however, and it took a long time for their scars to heal. Yet the lesson taught by the vigilantes of Virginia City was a good one. Only sporadic reports of incendiary attempts were made late into the century, and the Comstock was never again plagued with the kind of organized arson activity perpetrated by the "black sheep" of the fire department.

XIII. A Love-Hate Affair

One of the single largest supporters of the fire departments of Gold Hill and Virginia City was the railroad. Although its construction came several years after the founding of either fire department, the Virginia & Truckee Railroad always seemed to be one of the biggest contributors to the fire funds of both towns. In Gold Hill, the fledgling Yellow Jacket Hose Company No. 2 found the railroad willing to present a sizeable donation to the company for the purchase of helmets, hooks, ladders, hose, and a hose cart. In Virginia City, firemen always seemed to find donations for large blocks of tickets for the annual ball or other fund-raisers from railroad executives.

For many years, the Virginia & Truckee helped promote cooperation among the fire companies by offering reduced fares for special occasions and excursions. Sometimes ticket prices were reduced by more than 50 percent, to allow the firemen of Gold Hill to attend a formal dinner and ball hosted by Warren Engine Company No. 1 or Curry Engine Company No. 2, in Carson City (the usual fare was one dollar per person). Carson City traveled to Gold Hill and returned on June 5, 1891; Liberty Engine Company No. 1 attended the Warren Engine Company ball on June 17, 1891; Warren Engine Company returned to Gold Hill on October 28, 1892; and 215 firemen from the Carson City and Comstock fire departments attended a picnic at the Bowers Mansion, on a special train made up of an engine and four cars, on June 27, 1909.[1]

When the grand railroad celebration was held in Sacramento in 1869, *Gold Hill News* editor Alf Doten, who was also a member of Washoe Engine Company No. 4, wrote to Charles Crocker to request a special pass on the Central Pacific Railroad for the firemen and fire engines of the Comstock. Crocker responded quickly: "We have not passenger cars sufficient enough so you will have to come on flatcars. All firemen in uniform will be brought and returned free on a special train. Those not returning on the special train will be charged full fare on any other train. Make this fully known, so that no hard feelings will be engendered by mistake." A delegation of ten firemen from each of the companies in the Virginia Fire Department, one hundred members of Washoe Engine Company No. 4, sixty members from Liberty Engine Company No. 1 in Gold Hill, and representatives from the other Gold Hill companies boarded the train in Reno. They were resplendent in full dress uniforms, and hauled with them the hand engines of No. 4 and Liberty Engine Company No. 1, which happened to be the former Howard Engine No. 3 of the San Francisco Fire Department.[2]

The celebration was a spectacular one. The *Gold Hill News* reported that the returning firemen had had a good time, "and the goose hung high." It also approvingly noted the railroad's contribution of the special train on a Sunday, "although they run no trains on Sunday."[3]

When Good Will Hose Company No. 2, in Virginia City, purchased a hose carriage in Philadelphia in 1870, free freight and shipping was arranged on the transcontinental railroad. Although the carriage was lost for a time and misdelivered to Truckee, the members of No. 2 were still grateful for the complimentary shipping arranged by Virginia & Truckee Railroad official F. A. Tritle.[4]

The Virginia & Truckee steamed up a special train to haul the steam fire engines from Monumental Engine Company No. 6 and Yellow Jacket Engine Company No. 2, along with No. 2's hose carriage, to the great fire at the lumber and wood yards of the Carson and Lake Tahoe Wood, Lumber and Fluming Company, in 1877. The equipment boarded the special train at the Gold Hill depot. All hands and apparatus were aboard flatcars, and they hung on for dear life, with the engines secured by heavy ropes, as the locomotive and its precious cargo raced for the big light at the bottom of the canyon south of Carson City.[5]

Even though the railroads proved to be important benefactors of the Comstock fire companies, for Liberty Engine Company No. 1, the relationship was one of love-hate, occasionally approaching an outright shooting war. The trouble began in 1874, when a flatcar piled high with lumber lost its load on the curve behind Liberty's engine house, and a number of the planks flew through the air like spears. The result, the *Gold Hill News* said, was that the lumber struck with such force as to "create a number of ventilators in the side of the building." There was considerable grumbling from the firemen over the matter, but the railroad paid the cost of repairing the damage, although with a little grumbling of its own.[6]

The railroad was also blamed by Liberty's volunteers for the holes in the windows of the fire hall alongside the tracks, though it was commonly known that tramps and youngsters stealing rides on the daily trains to and from Virginia City were responsible for the rocks pitched through the glass. It was great sport in those days to test one's skill at pitching boulders through the small panes of glass from a moving train, and the Comstock produced more than one sharpshooter.

The weight and action of the daily trains passing between the Liberty engine house and the Miners' Union Hall in Gold Hill began to break down the large bulkhead behind the firehouse. After many months of grumbling, Liberty Engine Company No. 1 voted, in 1890, to have the secretary of the company write to railroad superintendent Henry M. Yerington and request that damage to the bulkhead be repaired. Yerington responded by ordering a crew to begin work on the damaged bulkhead, but the days of railroad cooperation with the fire department were quickly coming to an end, along with the days of profitable rail operation on the Comstock.[7]

Tunnel fires were especially exasperating to the railroad officials, and were equally troublesome for the fire department. From time to time virtually every tunnel on the railroad's Comstock line burned. When the American Flat tunnel burned, the railroad hauled the hand engine and hose carriage of Liberty No. 1 from Gold Hill to extinguish the fire. The hand pumper draughted water from the tender of the engine, no other supply of water being readily at hand. In 1902, the straight tunnel known as the Yellow Jacket tunnel was burned out. According to the diary of Alex M. Ardery, assistant superintendent of the railroad, the fire left only five or six sets of support timbers at the west end, which he said would have to be removed:[8]

> First alarm given at 1:30 p.m. by Eng. 12 at Virginia Roundhouse. Hose were laid from Divide & Gold Hill water pressure very poor and little could be done. By 3:30 fire consumed most of straight or Jacket Tunnel. Five timbers were destroyed. Supposed to have been caused by spark from passing engine. Last trains through were Extra McDoneling. #4 east bound at 8:30 a.m. returning west at 9:10. Passenger east bound at 12:10.
>
> All passengers and freight must be taken by teams from Gold Hill.[9]

The engine J. W. Bowker of the Virginia & Truckee Railroad was fitted with a special steam fire pump. The pump was powerful enough to throw a strong stream of water a great distance and was effective in extinguishing fires started by sparks along the railroad right-of-way. (Nevada State Museum)

A tally of the costs of the fire and rebuilding of the tunnel, kept by the railroad auditor's office, shows the Virginia & Truckee paid one hundred dollars to the "Va & G.H. Fire Deps.," for donation of services at tunnel fire." In addition, the cost sheet reveals a five-dollar charge from E. C. Swift for a team of horses to haul the hose cart to the tunnel. In all, the 1902 tunnel fire cost the railroad $7,604.02.

At eight a.m., November 17, 1903, nearly a year later, railroad superintendent Henry M. Yerington received the following telegram at the Palace Hotel in San Francisco: "Homestead tunnel Gold Hill discovered on fire four o'clock this morning too late for firemen do any good. total loss. Some caving near both ends, sent engine for local crew. Kirk will be at Gold Hill soon. A. M. Ardery."[10]

The fire was to prove costly not only in terms of revenue lost while the track was shut down during repairs, but also in repair and incidental costs, as well as the loss of friendly relations with the firemen of the Gold Hill Fire Department, and especially Liberty Engine Company No. 1. Ardery's diary of November 17, 1903, presents an interesting look at the situation from the railroad officer's point of view:

> For past ten days tunnel gang was putting on new zinc lining [to help prevent future fires]. At night or quitting time it was custom to leave a lantern on each side of any opening left—the engines of any train would shut off at such signals so as to prevent throwing sparks. This was done regularly—After train past it was the duty of the Gold Hill watchman to remove lanterns and give the Tunnel a careful examination several times during the night. This he claims to have done but no one for a moment believes him—Knowing nothing could be done to save loss started special [train] to Gold Hill for local train crew, so was able to make up local train from Gold Hill—began work both ends tunnel loading caved ground on flat cars 7 a.m. Nov. 18.

At least five box cars, three flatcars, a coach, and a locomotive—No. 17—were stranded in Virginia City by the fire, according to the diary. Repairs on the tunnel were estimated to require at least three months' work and 200,000 feet of lumber. What Ardery could not project was the controversy which would erupt between the railroad and the firemen of Gold Hill, which would last more months than it would take to repair the burned passage.[11]

During the fire, five lengths of hose were loaned to the railroad and left near the tunnel on a standby basis. During the night, the damp hose froze and was severely damaged. The firemen of Gold Hill wanted their hose replaced, but after three months the trustees of Liberty Engine Company No. 1 were still unable to schedule a meeting with representatives of the railroad company, even though they had promised to meet with the firemen and inspect the hose in question. One fireman suggested that the chief engineer should be censured for allowing the hose to be left on the Homestead Road for so long, but the chief told the firemen, "Mr. Kirk told him the railroad would pay for any damage occurring to the hose or property of the Engine Co." The chief was then instructed to see Kirk and advise him of the displeasure of the fire company and the apparent snub of the Virginia & Truckee. A meeting was held shortly after, and about a week later, Yerington received the following communication from his representative in Gold Hill:

> The Gold Hill Fire Co. has called my attention to five lengths of hose they claim has been made useless by freezing while in use from time to time while spiking was being driven in the tunnel.
>
> As the spiking progressed fire was uncovered from time to time, so arrangements were made with the fire chief—Valentine, to have the hose cart at the hydrant to have hose convenient at any moment.
>
> Valentine at this time was employed as a laborer cleaning out the tunnel.
>
> Yesterday afternoon the fire company trustees asked me to come in to the engine house and see the hose and I did so.
>
> They showed me three lots of hose viz:-
>
> 1 New hose that had been in service and not damaged.
>
> 2 The five lengths in question they claim was just as new but damaged by freezing.
>
> 3 The two lengths of old hose.
>
> I failed to see any comparison between that which they claimed to be new and not injured and the five lengths.
>
> The five lengths have a decidedly old appearance and an inferior coupling as compared with the good and I told them so.
>
> They replied no, that I am mistaken—that the hose is strained from freezing and can not anymore be counted upon to stand pressure.
>
> I asked them what they wanted, and they replied they thought it was no more than right for the railroad company to replace the injured hose.
>
> I told them I thought it looked like a jilt to work us for new hose; they said no not at all—well I said you write to Mr. Yerington and let him know what you want.

> It seems to me the company pay their taxes and are entitled to full fire protection regardless of what hose or labor are employed in giving it.
>
> The whole thing seems to me to be a cheeky piece of business but I feel that it might not be well to antagonize these fellows who work in time of fire.[12]

The company did write to Yerington—twice—but his response brought little satisfaction. According to the second letter to Yerington, from John Meagher, secretary of Liberty Engine Company No. 1:

> I am in receipt of yours of 20th inst. in which you express surprise at the contents of my letter. And while we acknowledged what you say in your communication to be true, we are satisfied you are not correctly informed in the matter. In the first place, this company does not ask pay of any person or corporation for any damage to their property incurred at a fire or while the same is in charge of this company. The fact of the matter is, that after this company had rendered what services it could during the fire, the hose instead of being reeled up and returned to the Engine House was left on the Homestead at the request of Mr. Kirk and Twoomy. The hose was loaned the railroad and was used by the men in your employ for about six weeks, part of the time it was left unreeled and full of water and I am informed it was frozen on three different occasions, so it required warm water and considerable pounding to break the ice so water could flow through it. The hose was the same which we used upon our jumper and was in good condition when left on the Homestead. When Mr. Kirk requested that the hose be left so it could be used when required, he informed the Chief of the Department that if there was any damage done the hose the R.R. Co. would pay for it.

In the view of the firemen, Kirk had been playing both ends to his advantage, attempting to keep in Yerington's good graces while stalling the Gold Hill firemen. Meagher warned the railroad superintendent that had a fire occurred anywhere in Gold Hill while the hose was on loan, the fire department would have been severely handicapped. He reported that the hose had been left only upon request of the railroad officials after the fire company had done its duty.[13]

Kirk was promptly called to answer the charge that he had requested use of the hose and had promised to pay for any damages to it. He admitted to Yerington, "It is true that I did request the chief of the fire department to let us have the hose for protection from fire which I could plainly see would be breaking out from time to time . . . " Kirk further stated that the fire chief had willingly loaned the hose and that the firemen had been "very accommodating in the matter." Although Kirk finally took responsibility for requesting the hose, something he had failed to tell Yerington in his first letter, he still found fault with the claims of the firemen. "The objection I made was that of the hose they showed me," he wrote. "I could not see that it does not have the appearance of having been as new as they claimed it was just before it was turned over to us for use and I could not see that it was injured by straining from freezing and told them so." Even with this admission, however, the situation did not charge. The minutes of Liberty Engine Company No. 1 for March 8, 1904, simply state that the secretary had yet to receive an answer from Yerington. "It was evident that the R.R. Co. would not pay anything for the damage done and as the Co. could do nothing more in the matter it was dropped."[14] Thus was born a bad feeling about the railroad that would last until the rails of the Virginia City line of the Virginia & Truckee were finally taken up by the scrappers in 1938, despite a few members of the engine company who were railroad employees.

XIV. Fire In The Mines

There was perhaps no fire more terrifying to the residents of the Comstock than one in the inky depths of the mines. While fire above ground might be controlled or stopped with relative ease, and little or no loss of life, this was not the case in the bowels of the earth beneath Virginia City and Gold Hill.

Time and again the shrill sound of the steam whistles at the mines and mills helped sound the alarm of fire throughout the town. But a long wail of the whistles meant only one thing—someone might be trapped at the 1,100-foot level by an inferno producing noxious smoke and deadly gas.

Too often, the firemen who met at the mouth of a burning shaft were virtually helpless to fight the fire. Sometimes their quick action prevented the destruction of a hoisting works or the spread of fire to the rest of the city. Often, however, low water pressure, inadequate supplies of hose and manpower, or the combination of heat, smoke, toxic gases, and occasionally a store of hidden explosives, worked against them. There was also the constant problem of distance from the hand pumpers at the shaft openings to the fires deep beneath the surface.

On July 21, 1869, the extensive hoisting works of the Chollar-Potosi Mining Company became a sea of seething, rolling flame. The minute the works were discovered to be on fire, an alarm was sounded throughout Virginia City. Every engine company, hose carriage, and the hook and ladder truck went to the scene, but there was a scarcity of water, and little could be done. The firemen of Young America Engine Company No. 2 stood with nozzle in hand, waiting for water to come while heat and smoke thrust out at them from the burning structures. At last a stream arrived from a cistern on the Divide, and only through the most strenuous efforts saved the mining company's boardinghouse and a number of smaller dwellings across the street.

The fire originated on the roof of the boiler room, where the boilers had been installed only two feet from the dry wooden roof, and close to the heated smokestack of the works. A new employee had searched in vain for a bucket of water to extinguish the blaze, despite the fifteen filled water buckets always kept on hand in the building. The mining company had organized its own force of men to work the company's steam pump in the event of fire; but the crew was too late to quell the flames that rolled swiftly through the hoisting room, boiler rooms, and adjacent buildings, within scarcely two minutes after they were discovered. The engineer on duty had only enough time to open the steam valves of the boilers, to prevent them from exploding, before he escaped the flames. The open valves caused a great hiss of escaping steam, adding to the pandemonium of the flames, the collapsing walls, and the shouts of firemen screaming for water.

There was no time to get the men out of the mine or even to send them a warning, but the miners working in the depths of the mine sensed the danger. When smoke began to pour down the air shaft into the mine they began scrambling for a way out. Four men on the 500- and 900-foot levels got out of the shaft before they were trapped, by climbing down the ladders in the pump shaft to the 1,100-foot level, then passing north through it into the 1,030-foot level of the adjacent Hale and Norcross Mine. The connection through which they escaped had been made only four or five weeks earlier.[1]

The miners working in the Belcher shaft in 1874 were not as lucky. A lighted candle left sticking in a timber set the mine on fire, and within minutes the shaft was filled with thick, dark clouds of smoke. Throughout both Gold Hill and Virginia City, the alarm bells sounded in chorus with the steam whistles, sending the fire departments of both towns out in full force. They were followed by a large crowd of spectators, including women and children sobbing hysterically at the thought that their husbands and fathers might perish in the flames.

Under the direction of the mine superintendent and foreman, the hoisting works hose, connected to the large steam pump inside, was aimed at the mouth of the shaft. But within one-half hour after the water began pouring down into the mine, the flames burst up with a terrific roar, several hundred feet into the air, hurling fragments of rock in every direction. The fire resembled a volcano venting its fury.

Although the *Gold Hill News* initially reported that all of the men working in the shaft had narrowly escaped death, in fact at least four men were badly burned and another killed. Patrick Kelly, one of the men who volunteered to go into the mine to try and stop the flames, was badly scorched by a sheet of fire which burst upon him in the air shaft. He wandered off in confusion and shock, and fell about twenty-five feet down a chute and into an incline, where he died. Desire Cornellier was severely burned from the hips upward; Frank LeClare was slightly burned, but suffered scorched lungs from inhaling hot air and flames; Richard Pollard was burned from his hips to his head; and William Johns was severely burned on the chest. All the doctors could do to relieve the suffering of those men was to cover their burned skin with lime water, linseed oil, and lint, as a temporary dressing.

The fire moved with such rapidity that all efforts to extinguish it were useless. There was no time to hoist the four miners working below to the surface. "It was terrible to be forced to abandon the men in the mine to what even those hopeful feared must be a frightful death," the newspaper reported. But that was the only alternative. All that could be done on the surface was to pray for the men trapped below. Fortunately, there was a means of escape for them through the connection between the old shaft and the new workings. "The experience of the men, who were down in the shaft, is one that they will not want to repeat during the rest of their lives," the newspaper summed up. "It may be that they knew there was no danger of the flames reaching them, but if not they must know how it feels to face a terrible death."[2]

The explosion of a coal oil lamp at the Utah hoisting works north of Virginia City, just three weeks prior to the Great Fire of 1875, sent firemen and equipment rushing along the streets towards Geiger Grade and the bright orange light in the sky. Only Monumental Engine Company No. 6, from the Divide south of Virginia City, made it all the way to the fire, leaving the other companies behind on the steep, rocky road to the mill. Even then, the Monumentals were too late in bringing their steam fire engine and one thousand feet of hose to the fire, and the works were destroyed. "They deserve credit for the spirit they displayed," one newspaper suggested.

The alarm had first been sounded by May Bassett, a young lady living near the mine. She stood at her post and continued blowing the steam whistle at the mine until the heat and flames drove her off. Robert Laird, one of the employees at the mine, descended an old shaft to determine the fate of the men working in the main shaft below. After getting down about one hundred feet on old ladders, he was hailed by them from the bottom of the main shaft. The good news that they were alive was relayed to the surface, where a shout went up that echoed in the nearby canyon. A rope was lowered and the men were hauled to the surface, all uninjured. The hoisting works and machinery of the Utah Company were a complete loss, however, with considerable damage to several adjacent buildings and much of the equipment. In all, the Utah reported a loss of $70,000, but not one life.[3]

Such was not the case on April 7, 1869, however, when the greatest mining calamity to occur on the Comstock Lode began in a fire on the 800-foot level of the Yellow Jacket Mine, near the shaft of the Kentuck, in Gold Hill. The fire spread rapidly, sending a column of thick black smoke into the air and causing an extensive fire alarm to be sounded in Gold Hill and neighboring Virginia City. "It was the hour when the shifts were changing, so that fewer men were at work in the mines than usual, and this chance saved many lives," Eliot Lord wrote of the disaster in 1883.

> John Murphy, station-man at the 800-foot level of the Yellow Jacket shaft, heard a sound like a gust of wind roaring through the drift, and saw the fifteen lights of the station at once extinguished. The foul blast stifled him, and he crouched on the floor, wrapping his rubber coat about his face. In a moment he lost consciousness, but could remember when rescued that he heard a pitiful cry come up the shaft from a lower level: "Murphy, send me a cage; I am suffocating to death!" Two miners at work in the 800-foot level of the Kentuck heard a gale-like roaring through the drift and were instantly overwhelmed by its fierce blast of smoke and gas. One struggled through the stifling atmosphere to the Crown Point shaft and was saved; the other fell dying in the drift beyond hope of rescue.[4]

In the adjacent Crown Point Mine, forty-five men had just been lowered on the cage when the blast of smoke and gas blew through. Gasping for air, they rushed the cage, realizing it would be their only hope for survival. The scene was one of panic, according to Lord's account:

> Three brothers, from Yorkshire, England, all strong young men, were working in the Crown Point Mine. George Bickle, one of the three, stood on the cage insensible, leaning over his dead brother, Richard, and holding him with a grip which could scarcely be loosened. Richard Bickle had sunk down upon the bottom of the cage as it was drawn up, and his head and arm were torn almost

completely from his body by the side timbers of the shaft. The dead and dying brothers were parted by kind hands, and George Bickle was laid tenderly upon a rude couch in the hoisting works by the side of the other sufferers, who, like him, were past all help of medicine. They gasped faintly for some hours and their troubled breathing ceased forever. The cage was lowered again to the lower levels of the mine, but no answering signal was given.[5]

Huge columns of thick, black smoke began to rise from the Kentuck, Yellow Jacket, and Crown Point shafts; the entire Comstock knew it was facing its most terrible and painful calamity. Relatives of the miners began to gather about the shaft, women silently weeping, children not fully comprehending the crisis before them, as the giant priest from St. Mary's in the Mountains, Rev. Patrick Manogue, moved about trying to comfort them.[6]

All of the men working in the mine, down to the 800-foot level, were trapped by the suffocating smoke, toxic fumes, and sheets of flame racing through the deep shafts and air shafts in the three mines. The blaze was fanned by the ventilation caused by the connecting drifts and air shafts in the three mines. The smoke and fire worked its way from the Yellow Jacket into the Kentuck, and then into the Crown Point, forcing the miners in those claims to make a hasty exodus. A cage was lowered into the depths as onlookers and the families of the trapped miners gathered in front of the hoisting works. From the 900-foot level came a cargo of dead men, killed by the thick smoke as they made their way to the surface. From the 700-foot level two more victims were brought to the surface.[7]

Every effort was made to rescue the trapped men. A strong draft began to move down the Yellow Jacket shaft and up the Crown Point, clearing away much of the dense smoke and allowing small search parties of firemen and miners to go into the burning shaft. They entered repeatedly, but with little success. They found four victims on the 900-foot level of the Yellow Jacket, and two more on the 700-foot level of the Kentuck. Thick poisonous smoke continued to pour up the Crown Point, but because the blower of the mine was kept running throughout the fire, it was thought that the fresh air pumped into the lower levels might get below the smoke, and there might still be men alive. A cage was sent down the shaft to the 1,000-foot level, with a lighted lantern and a message reading, "We are fast subduing the fire. It is death to attempt to come up from where you are. We shall get you out soon. The gas in the shaft is terrible and produces sure and speedy death. Write a word to us and send it up on the cage, and let us know where you are." The hopeful waited for a return signal, but were met only with silence. After a few minutes, the cage was raised, note intact, the flame of the lamp extinguished from the lack of oxygen in the mine's depths. The shifts of the miners were then checked; one missing from the Yellow Jacket, four unaccounted for in the Kentuck, twenty-three missing in the Crown Point.

At midnight, it was decided to attempt to penetrate the hell below, through the 900-foot level of the Yellow Jacket, then into the lower depths of the Crown Point. Lord wrote of the rescue mission, "The days of chivalry are doubtless past, yet to penetrate into the depths of a burning mine and brave the imminent perils of suffocation by smoke and gas, on a forlorn hope at best, is a deed which few knights would have sought to match." The knights who answered the call on that dark Friday were all firemen. A. A. Putnam and Bill Lee, both of Liberty Engine Company No. 1; Richard Mercer, of Yellow Jacket Engine Company No. 2; and Henry Aine, of Nevada Hook and Ladder Company No. 1, moved towards the shaft and possible death, into one of the most gruesome adventures of their lives. They made their way to the 1,000-foot level of the Crown Point and groped their way through the choking smoke and gases of the drifts. What they saw sickened them. Lord wrote:

> Dead men were lying on the floor of the level as they fell in agony of suffocation, with their mouths glued to cracks in the planks or raised over winzes, turning everywhere for one last breath of fresh air. Their faces were flushed and swollen, but the features of well-known friends were not past recognition. Farther on, however, in the well at the bottom of the shaft, frightfully mangled bodies were found of wretched men who had met an instant death in their wild instinct to escape from torture. One poor sufferer had climbed up the shaft to a point between the 800- and 900-foot levels, where he was found hanging to the ladder with one leg fast inside, and still clasping the rounds with so firm a death grip that he could only be plucked away by force.[8]

The brave firemen worked quickly to secure the bodies to planks and hoist them to the surface. Above ground, the shaft was roped off to prevent a rush of grieving relatives that might hamper the operation. As the victims were brought up, a pitiful cry was heard again and again: "My God! Who is it this time?" Some of the bodies were so mangled or decomposed from the heat and steam that efforts were made to hide them from relatives. It was one of the most agonizing nights in the history

It fell to volunteer firemen to attempt rescues during fires in the mines. Members of Liberty Hose Company No. 1 volunteered to descend into the burning mines on "Black Friday" in 1869. (Warren Engine Company No. 1)

of the Comstock, and one which would live in the memory of all present that day for the rest of their lives.[9] The funerals that followed were among the most extensive ever held in Gold Hill or Virginia City. Long lines of mourners accompanied the bodies to the various cemeteries in Virginia City and Gold Hill. The military and fire companies turned out in full dress uniform to say farewell to friends and relatives.[10]

The fires in the three Gold Hill mines continued to burn. Despite the efforts of the fire companies to extinguish the blazes by pouring water down the shafts, thick columns of smoke continued to come up through the Crown Point. Steam was forced into the mines to try to quell the blazes, but it succeeded only in clearing the air of some of the foul smoke and gases. Small search and work parties began to explore some of the drifts, finding three more victims. As air began to work into the mines, smouldering fire brands were fanned from hiding places, and those in the drifts worked as hard as they could just to keep the precious ground they had won from being claimed by a new fire.

The mines were finally closed in one last effort to smother the flames, but were reopened several days later to allow an assessment of the damage and to begin work on closing the passages which connected the mines and created the strong drafts that had fanned the fires. Some of the galleries were closed completely, entombing yet undiscovered bodies. Six months later, miners still occasionally reported driving picks into smouldering brands, and several men were overcome by a sudden rush of gas while working in a stope between the 600- and 700-foot levels. Fortunately, they all recovered upon being brought to the surface and fresh air.[11] Much of the damage caused to the mines by this fire was never repaired, and some of the closed galleries were never worked again. The fire sounded the death knell of forty-five men, and created one of the darkest pages in the history of mining in Nevada.

XV. Conflagrations

Virginia City and Gold Hill were plagued with major fires almost from the beginning of their existence. In virtually every case, something hampered the efforts of the firemen to battle back the "fire fiend." Scarcity of water at the location of a fire, inadequate supplies of hose, and the damaging effects of wind fanning fire, all worked against the efforts of the volunteer firemen of the Comstock. When that happened, the newspapers exploited the situation fully by reporting the stories with as much detail as possible and using the biggest and boldest type for headlines to describe the "Extensive Conflagration," the "Great Fire," or how the "Fire Fiend Strikes Again."

It was on such occasions that the community was suddenly awakened to the reality that the town was fair game for destruction by fire at any time, despite the best efforts of the firemen and their machines. Through the years, major fires consumed whole blocks of buildings, destroying entire business districts or residential sections. A sampling of the fires suffered on the Comstock fails to begin to describe the serious situation that faced Virginia City and Gold Hill time and again, as the fire bells pealed, the steam whistles shrieked, and the volunteers donned their leather helmets.

In 1864, an entire block was destroyed on the Divide between Virginia City and Gold Hill. Losses included the Golden Eagle Hotel, where the fire originated, the Eureka Saloon, C. W. Larrowe's blacksmith shop, the Mount Davidson House, and a Chinese washhouse. In addition, both the Junction House and the Miners' League were badly scorched by the flames. The loss totaled more than $16,000. None of the buildings was insured, putting virtually everyone out of business and out of luck.

The fire, which broke out early in the morning, in the rear of the hotel, was reportedly the result of a careless miner smoking in bed. The flames spread to adjacent buildings almost immediately. The firemen struggled against the blaze as best they could, but faced an almost insurmountable obstacle. There was little water near the business block, and not enough hose to supply the engines at the scene. Only when the buildings on the west side of the street caught fire could the firemen get an effective stream of water on the fire. Knickerbocker No. 5 finally managed to find a supply of water and began pumping furiously, while Eagle No. 3 prepared to receive a relayed supply of water from the engine of Young America Engine Company No. 2.

Virginia Engine Company No. 1 attempted to dash through the fire with its engine to reach a reservoir just south of the fire. Nearly every man had his clothes almost burned from his back, and all suffered from the heat and flames as they passed. Those with No. 1's hose carriage suffered even worse, passing through the sea of flames with the rope burning furiously in their hands. They were forced to abandon the hose carriage or be burned to death at one point, but Billy Warnock and several others rushed forward and drew the carriage to safety. Two lengths of hose were burned, and one side of Virginia Engine No. 1 was stripped of its paint. Somehow the firemen prevailed, and within a short time they beat back the fire and saved the entire block on the west side of the street, opposite the buildings where the fire had originated.[1]

Lizzie Williams, the young woman who occasionally manned the brakes of the engines and often went about recruiting bystanders to help at fires, was the cause of a lively but destructive Virginia City fire in 1869. Lizzie was living in a basement room in A. C. Scott's American Exchange on C Street, next to the Wells, Fargo building. On the evening of March 12, the young fire buff accidentally knocked over a kerosene lamp in her room, spreading flames across the tinder-dry wooden floor of her apartment. The building was of wood, extremely dry and ripe for a fire. Within five minutes, flames had reached the roof and were bursting through the front of the building.

Scott, a member of Young America Engine Company No. 2, his family, and other occupants of the building had scarcely enough time to evacuate the inferno. Most of them reached the safety of the street wearing underclothes or wrapped in blankets. Men working at the nearby Gould & Curry Mill got the first stream on the fire with the mine hose. Assisted by Knickerbocker Engine Company No. 5, they were able to save several adjacent buildings. Young America Engine Company No.

2, whose engine house was just down the street, moved quickly into position and got the next stream on the fire, followed by Washoe Engine Company No. 4 and Monumental Engine Company No. 6, from the Divide. The latter made the run from their hall south of town faster than anyone at the scene had expected.

Despite the marshaled efforts of the volunteer firemen and their hand pumpers, Scott's building, Battu's paint shop, and an adjacent restaurant were all destroyed. Charley Scholl's gunsmith shop was badly scorched, and some stock was damaged when it was removed from the structure, but Scholl nevertheless opened for business the next day. The roof of the Wells, Fargo & Company building was damaged by fire and water, but the building and its contents were safe and ready for business the next day. All of the buildings on the D Street side of the fire had been saved by the hard work of the firemen of Virginia City and Gold Hill. Of the three buildings burned, not a single one carried insurance to cover the combined $20,000 in damages.[2]

The damage was doubled on October 22, 1870, when nine buildings, valued at $40,000, burned. Had not a snow fallen the evening before, Virginia City might well have been completely destroyed as the flames leaped along their path of destruction. The fire broke out about 1:30 a.m., in the rear basement of the Washoe Stable on South C Street. Within one-half hour, all of the adjacent buildings were blazing, frustrating efforts to check it. The fire moved so quickly that the firemen had little chance to stop it before it burned through to D Street, although they were aided by a strong wind which kept the spread of the flames across C Street.

Young America Engine Company No. 2 moved into position quickly and "got first water" on the houses on the west side of C Street, which were smoking and in danger of bursting into flames. Washoe Engine Company No. 4 rolled into position next, taking water from the cistern opposite Smith Street and preventing the spread of fire past its position. Monumental Engine Company No. 6 from the Divide soon joined the battle, along with Knickerbocker Engine Company No. 5, which stationed its engine at a cistern near the Virginia House, where the heat was reported hottest. No. 5's firemen continued to play water on the inferno despite the heat, pumping the cistern dry before moving back to another source of water.

It was to be one of the more lively fires attended by the Comstock firemen. During the heat of the battle, a fireman named Wylie was caught in the front parlor of the Virginia House as it fell down a bank with a loud crash, carrying him with it amid flame and burning lumber. Fortunately, the young man rose to his feet and called with a wave of his hat for a stream of water to protect him until helpful hands could rescue him from what seemed a certain fiery tomb. After the fire had been put out, too much whiskey caused some fights and led to the incarceration of a few firemen by Chief of Police George Downey, also a member of Nevada Hook and Ladder Company No. 1. While escorting one of the party-goers to jail, Officer Higbee was shot in his left calf. The culprit was whisked off with a new fervor. When the smoke had cleared, and hangovers ensued, it was found that only four of the fire victims had insurance, amounting to $14,200, against a tally of losses amounting to $40,000.[3]

The year 1871 proved to be a hard one for the Comstock, with two devastating fires striking only a few months apart. On February 1, about 1:20 a.m., a fire broke out in a building behind the old Merchants' Exchange, which was being used at the time as a lodging house. Within minutes the flames were leaping from building to building, aided by a strong wind, and there was little anyone could do. The heat was so intense that it left the iron doors at the rear of the post office warped, and scorched the wooden doors inside. At least forty buildings were destroyed between Union Street on the south, Sutton Avenue on the north, C Street on the west, and E Street on the east. Piper's Opera House was saved only through the efforts of a small band of firemen who made a stand in the middle of the street, virtually enveloped in flames, and poured water on McKay's stables to keep the flames away from the opera house.

It was to be a deadly fire as well as a destructive one. Two bodies were found in the smouldering embers of Mrs. Sherman's lodging house, both burned virtually beyond recognition. Hours later, only one victim could be identified, through a watch found on his body.

The report of yet another death luckly proved false. Former Sheriff Cummings, a member of Washoe Engine Company No. 4, moved about the burning buildings near Piper's Opera House, taking the nozzle in hand and leading the fight to prevent the spread of the fire and destruction of the opera house. Several times he dashed towards the flames with the hose, only to be lost in the smoke and flames around him. At one point he was reported to have been killed in a burning building, but, unknown to his fellow firemen, the former lawman had moved through the building's back entrance, leaving the hose and nozzle behind when the heat became too great. When his fellow firemen hauled

Extent of destruction from the Virginia City fire of August 20, 1871.

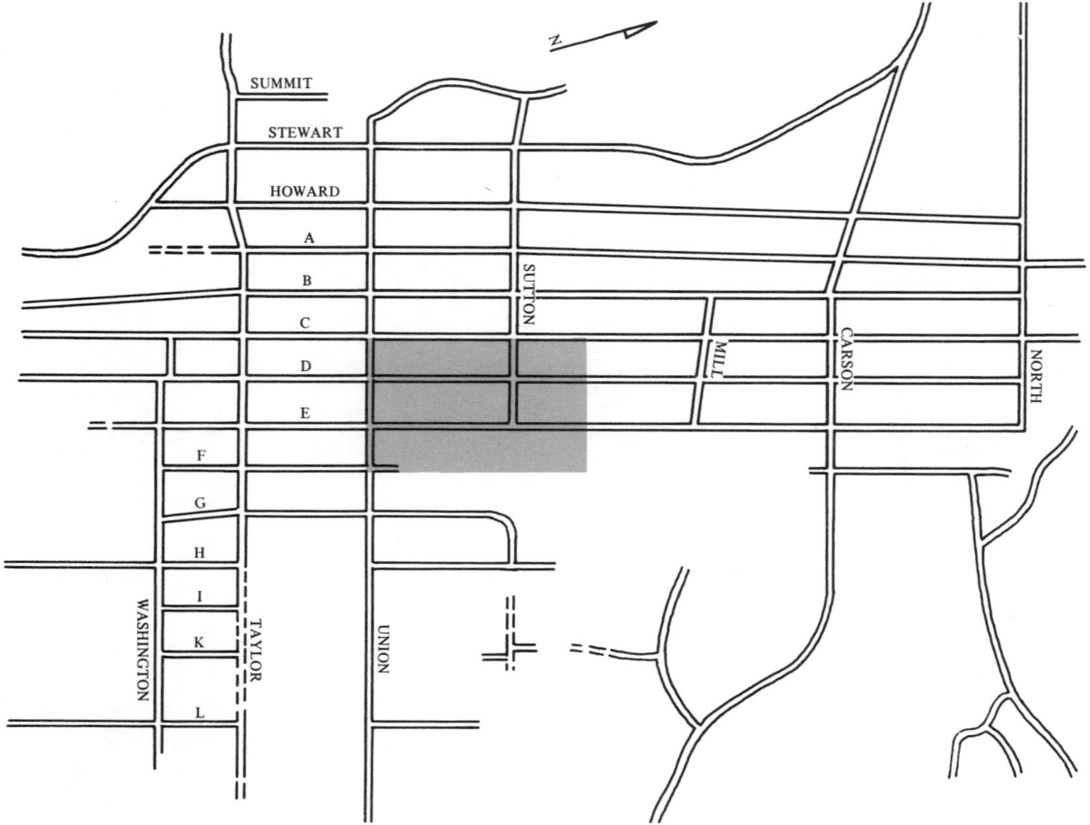

the line back through the building without him, they became convinced of his demise.[4]

In August, virtually the same district was again levelled by fire, except that it did not burn through to C Street this time. It did, however, extend one block further east, to F Street. The fire broke out at about 3:30 a.m., making the response time from warm beds a bit slow. When the firemen finally assembled, they found that someone had disabled Young America Engine No. 2, so it was unable to render any service. The men at the Virginia Consolidated Mining Company checked the advance of the fire in their direction through the use of the company's steam engine and hose. Only a small portion of the furniture in the houses of the city's red-light district was saved, and the occupants added a colorful sight to the pandemonium as they rushed from their small cribs in night clothes, blankets, and little else.

Almost all of the buildings destroyed were in the process of being rebuilt after their destruction in the February fire. Although a few were insured, the fire's destruction proved almost complete. Its psychological damage was also great, with businesses of long standing giving up, and disgusted proprietors selling their lots before smouldering ruins had cooled. There were no deaths during the blaze, but Willie Taylor of Neptune Hose Company No. 5 was badly burned as he attempted to help save furniture from one of the burning buildings. He found himself suddenly hemmed in, and had to be pulled from the blaze by his fellow firemen.

The fire proved to be the work of an arsonist. It was started beneath a flight of steps near the sidewalk between the house known as "The Three Bridges," and the house just south of it. Shortly afterwards, a man named Heilshorn was arrested on suspicion of setting the fire. He was tried and convicted of the arson, and sentenced to the Nevada State Prison in Carson City, where he served fourteen years.[5]

In 1873, five buildings were destroyed on the Divide, largely due to a lack of water on the hillside where the fire occurred. Despite a quantity of water in the town reservoir, there was no way to get enough of it up the hill to save any of the structures. The firemen of Gold Hill arrived to assist the Virginia Fire Department, and even Liberty Engine Company No. 1 hauled its heavy hand engine up the grade from Gold Hill to help fight the blaze; but there was little that could be done. Even the combined hose of Monumental Engine Company No. 6 and Neptune Hose Company No. 5 could not reach the fire from the nearest water at the Imperial hoisting works. As a result, several buildings were destroyed, and at least one fireman was seriously burned.[6]

Although 1871 had been a bad year for fires on the Comstock, 1875 was to prove the most devastating of all years in Virginia City. On May 19, an arsonist set fire to a frame building on B Street, between the International Hotel and the Masonic Building. The fire destroyed the Masonic lodge and most of the furnishings, and records of Escurial Lodge No. 7, Virginia No. 3, and Virginia Chapter No. 3 of R.A.M., as well as the DeWitt Clinton Commandery No. 1 of the Knights Templar.

The firemen paid special attention to the walls of nearby buildings that were exposed to the fire, and were able to prevent the spread of the blaze by wetting down the adjacent buildings and piles of lumber. J. W. Baker of Young America Engine Company No. 2 received a severe gash on the forehead from a nozzle that became unmanageable and was flipping around the street like a crazed snake. More serious injuries could have been reported, however. During the excitement, Peter Wilmot fell into an open cistern nearby, and would probably have drowned had not Martin Wiley gone in after him. As it turned out, both nearly drowned, and when pulled out of the cistern, Wiley was the worst off of the two.[7]

On June 29, two fires broke out during the afternoon. The first happened about one p.m., just north of the Episcopal church, and resulted from a defective stovepipe. It was quelled quickly with a few buckets of water. The second fire occurred one-half hour later, in the rear of the City Bakery, on B Street. Only through the quick work of Eagle Engine Company No. 3, and their new Babcock chemical extinguisher, dubbed the "soda bottle" by most of the town, was the fire prevented from spreading throughout the bakery and into adjacent buildings. A month earlier, the building adjacent had been fired by an incendiary, which caused speculation that he had returned to burn the bakery. The cause of the fire was traced to the bakery furnace, however, which had thrown sparks that ignited some woodwork.[8]

On September 3, the stove in the tailor shop near the Odd Fellows' Hall was blamed for the fire which burned the lodge, the livery stable of the Ferrend estate, and the building leased to George Schwab as a saloon. A number of other lodges had also been using the Odd Fellows' Hall, and all lost precious vestments and furnishings in the blaze. The National Guard saved its arms, but all uniforms were either destroyed by the fire or ruined by water. Battery A was able to save all of its belongings, however, with no damage reported. The Western Union Telegraph Company saved nothing but two machines and its books and papers. The

Independent Order of Red Men saved all their belongings, while the Odd Fellows' library, containing about 4,500 books, was totally destroyed. The livery stable housed numerous buggies and wagons, ten sets of harness, seven tons of hay, and five tons of barley and oats, all of which were burned.

The firemen found their task a hard one, due to the dryness of the season and the lack of sufficient water in the cisterns. The fire companies were forced to move from one cistern to another as the supply of water in each was exhausted. The Babcock "soda bottle" kept the buildings on the opposite side of the street from bursting into flame, by putting out spot fires as they occurred. After the fire was out, however, the firemen of the city had a lively time jumping from ruin to ruin, building to building, as the wind blew cinders and hot coals in showers on all the houses east of the burned district. It was a long night for the men in the leather helmets.[9]

Three weeks later, four buildings were destroyed when the Occidental Lodging House on C Street was set on fire by an arsonist. The four steam fire engines of the Virginia Fire Department were on hand, and within nine minutes each had a strong stream of water playing upon the blaze. But before forty minutes had passed, the water began to give out, forcing the engines to move back to more plentiful water supplies. Each time they moved back, another building was consumed by the hungry flames, prompting the newspapers to comment the next day:

> Had the supply of water been anything like adequate, the fire would have been extinguished in fifty minutes. And at least one quite valuable house saved. During all the time the fire was raging, and when the least delay was suicidal, not a drop of water was running into the cistern at which No. 2 was stationed, although the cistern is furnished with 2-inch supply pipes.[10]

This fire left a number of men with serious burns, and many eye injuries from cinders, dust, and smoke. It also left a controversy between the newspapers, the fire department, and the water company over the supply of water available to fight the fire. These injuries and the insults over water supplies would come to seem rather insignificant in the wake of the Great Fire of 1875, which would strike the city in only twenty days.

XVI. Virginia City's Great Fire of 1875

It was my fate to be in the middle of flaming San Francisco in 1906 and the great Berkeley fire in 1923 in which eight hundred houses burned, but the intensity of that mining camp blaze lingers in my memory as the fiercest of all. Mount Davidson seemed bursting in volcanic eruption." So wrote Wells Drury of Virginia City's Great Fire of 1875 in his book, *An Editor on the Comstock Lode.*[1]

In the year 1875, the Comstock world turned upside down. In January, the bottom fell out of the stock market. The *San Francisco Bulletin* reported that ten leading Comstock mines depreciated $17,814,800 in just twenty-four hours. The California Consolidated lost $70 a share, the Consolidated Virginia $30, and the Ophir $40. In February, the market hit bottom and began a very sluggish movement upwards toward recovery.[2]

In addition to this financial disaster, there had been substantial overestimates of the ore reserves on the Comstock, mostly for stock promotion purposes, which led many to invest heavily in mining stocks only to find that ore bodies were suddenly playing out. On August 26, 1875, the war between the so-called "Bonanza crowd" of mine owners and the "Bank crowd" of investors and bankers came to a head with the failure of the Bank of California. A short time later its leader, William C. Ralston, took a swim in San Francisco Bay that resulted in his death—enemies said it was suicide, friends called the death an accident caused by a stroke. No matter which, it seemed as if no further disasters could hit the Comstock Lode—until October.[3]

October of 1875 was to prove one of the most fateful months in the history of Virginia City. No fewer than a dozen major fires were reported in 1875; and in October, two major fires struck only two weeks apart, resulting in damages of more than $125,000 and the destruction of several of the town's major structures and business establishments.[4]

In the midst of this turmoil came a scandal involving the default of the treasurer of the Virginia Fire Department, who had appropriated $593 from the treasury for his own use. It seemed to the citizens of Virginia City that there was no end to the downslide of the fortunes of the mountain mining camp. The only bright spot in this otherwise dismal year was the visit of the famous Civil War and Indian campaign hero, General Phil Sheridan. The general, on tour throughout the United States, brought his entourage to the famous Comstock Lode and was greeted in traditional Comstock fashion with a large parade that included the city's volunteer firemen dressed in their red shirts and leather helmets and pulling their equipment. Following those festivities, a grand reception for the group was held at the International Hotel. Only one and a half weeks later. October 26, the beautiful International Hotel would be a pile of rubble.[5]

"The fate so long dreaded by the people of Virginia has come at last," the *Gold Hill News* sadly announced. "Two-thirds of the city is now a mass of smouldering ruins, the principal business part of the city, containing nearly all the valuable buildings, was in the space of a few short hours, razed to the ground." The fire began at about 5:15 a.m., October 26, 1875, in a small one-story lodging house on A Street, behind the Storey County Courthouse. According to later accounts, the fire was the result of a drunken row in the house "kept by a woman named Kate Shay, well but not favorably known as 'Crazy Kate.' "[6] Dan De Quille [William Wright] recalled the fire in his book, *The Big Bonanza*:

> The fire started at an hour when few persons were abroad. Only the butchers, bakers, marketmen, and other early risers were astir. The "owls" of the city, birds of prey that haunt the place all night, had disappeared with the gray of dawn and were in their first deep sleep; the time was an hour too early for the change of shifts in the mines; therefore at no other time, day or night, could the streets have been found more completely deserted.
>
> When the first fire bells rang, few persons heeded even though they heard them. Soon, however, the mournful and long-drawn wail of one steam whistle after another, in quick succession, was heard to join in sounding the alarm till the fierce clangor of the bells was almost drowned. The bells, loudly as they rang, only said: "There is a fire, and a great and dangerous one!" In the sounding of

> the whistles it was to be noted that there was no hesitation or timidity anywhere shown; each engineer pulled open the valve of his whistle to its full extent, at the first grasp of his hand.[7]

Although most of the town was slow in waking to the alarm, the four-wheel Babcock chemical extinguisher of Eagle Engine Company No. 3, and the hand engine of Washoe Engine Company No. 4 were both quickly on hand, their engine houses being only a block from the blaze. Both immediately went into action, but a fierce west wind carried flames and burning cinders for hundreds of feet. The efforts of No. 3 and No. 4 were to no avail, as the fire moved rapidly from one building to another before the other hand engines and the steamers of the fire department could get into position to begin their battle against the demon blaze. Absent from the battery of steam fire engines that day was Washoe No. 4, down for repairs at a foundry on the Divide. Even if it had been in service and ready to respond to the blaze at the outset, however, it is doubtful that it could have done much to prevent the holocaust. As the firemen moved their steamers and hand pumpers into position, the Babcock and No. 4's hand engine were forced to pull back to new locations.[8]

The new positions established by the fire companies did not last for long. The fire rapidly advanced along A and B streets, destroying the engine house of Eagle Engine Company No. 3 and threatening the courthouse and jail. Official records were locked in the "fireproof" vaults of the courthouse, and the prisoners were taken from the jail to the office of Judge Knox, down the street. But the flames moved swiftly and soon threatened that building, so the prisoners were again moved, this time to City Hall, on C Street, near Sutton Avenue. Next, they were transferred to the Union Consolidated Tunnel on the north end of town, and finally, to the Pacific Brewery on Geiger Grade, where they were placed in the basement beneath the saloon. Among the prisoners was Peter Larkin, accused of the murder of Daniel Corcoran, a member of Monumental Engine Company No. 6. When a fight broke out in the saloon overhead, Larkin made his escape through a small window in the makeshift basement prison. He was later captured in Seven Mile Canyon and returned to the custody of the sheriff.[9]

The wind was driving the flames at a tremendous rate, consuming building after building with such rapidity that the firemen were lucky to keep ahead of the flames with their lives. As De Quille wrote:

> Almost instantly the column of fire that was at first seen to arise began to assume the form of a pyramid. The base of this pyramid

The Babcock chemical engine of Eagle Engine Company No. 3 was the first apparatus on the fire ground in Virginia City on October 26, 1875, but it was ineffective against a large fire. Later it was repositioned near St. Mary's in the Mountains, but it was cut off and abandoned and was totally destroyed.

> rapidly extended into the sides of houses in all directions — the glass falling in showers from the windows to give ingress to the flames — structure after structure burst out in sheets of fire more rapidly than could be counted or noted down.[10]

Alex Daley was a youngster at the time of the Great Fire of 1875. His father, George, was a typesetter at the *Territorial Enterprise,* and was working on the morning when the fire broke out. In 1954, at the age of eighty-two. Daley recalled:

> At the time of the big fire, my father was trapped by the flames but passed the [printing] forms out of the cellar window, was hauled out himself, and took the forms to Gold Hill and printed his *Enterprise* there, partly on wrapping paper obtained at the drug store.
>
> We lived in a house up Mt. Davidson then and I saw the blazing city from a cliff nearby.
>
> An English couple, the Obistons, (who had the only grand piano in town) packed hastily and tried to drag their luggage up to our

> house but were overtaken by the flames at C Street, and had to run with what little they could carry—we had a family camped in every room of our house.[11]

Ferdinand Beck was a clerk at Fredericks' General Store at the time of the fire. At age eighty-two, he recalled the turmoil caused for him and his employer: "I happened to be in the store and doing the morning cleaning as the alarm came. I ran up the street and noticed thick smoke," he wrote. Joseph Fredericks was in the store at the same time, busy putting the cash from his register, the store's books, and a few other valuables in a large wicker basket, which he instructed Beck to take to the nuns at St. Mary's in the Mountains Catholic Church. When Beck returned, Fredericks ran off to join the other members of Knickerbocker Engine Company No. 5, who were battling the blaze: "The water gave out later in the day and the people worked frantic [*sic*] to save some of the houses. I stood on the roof which had shingles on next door from the store to keep it wet keeping the sparks off and prevent it from burning." Beck's house of employment was spared, but hundreds of other commercial establishments had been reduced to piles of ash and rubble. Had Fredericks' caught fire, Beck wrote, the whole block could have been wiped out from the oil and other combustible stock on hand: "The fire burnt all day and luckily it started to rain—but it came too late, the damage had been done."

Beck believed that his home had been saved from fiery destruction, so he stayed at the store to prevent it from being levelled. Walking home late in the afternoon, Beck discovered that the fire had reached his house as well. Sitting in front with some of their salvaged belongings was Beck's wife, who had tried to move their cast-iron cooking stove from the house in an effort to save it, much to his surprise. The only furniture that was saved was their couch. Even the large stack of wood he had gathered for winter only a few days before had been reduced to ashes.[12]

As the fire raged through the town, walls began to fall. John Scullion, of Monumental Engine Company No. 6, survived being buried beneath the crumbling walls of the International Hotel. Mike Malone, of Gold Hill, was killed by a falling wall on Union Street between D and C streets; and James Ketton, also of Gold Hill, was killed in the collapse of the walls of the Fredericksburg Brewery. One act of bravery saved the lives of two children. While climbing a burning stairway in a building on B Street, O. D. Williams discovered two sleeping children, apparently forgotten and left to face their fate. Williams took one under each arm and moved out of the burning building, and then left them in the care of Pat Ennis of the Emmet Guard.[13]

More and more men volunteered to work the brakes of the hand engines, pull hose, and otherwise assist the firemen of the city. Members of the Gold Hill Fire Department responded to the emergency with their hose carriages and the steamer of Yellow Jacket Engine Company No. 2. As the engine was put into position and began to pump water, however, a check valve blew off. This put the steamer out of commission for twenty minutes, while another valve was located and put into place. Valuable time had been lost, but, as the *Gold Hill News* reported:

> The fight against the fire was as vigorous as could be made, no effort being spared to stay its progress. The aid of powder was called in, and buildings blown up at different points toward the north and east of the city, but the flames leaped over the obstacles, and places which one minute were hundreds of feet from the fire were the next a sheet of flames. It appeared for a long time as if the new C&C hoisting works, down in Chinatown, were doomed, but a desperate and successful effort was made to save them.

In the midst of the flames and confusion, the crew of the Western Union Telegraph Company moved their records, a key, and a battery from the offices before they, too, were consumed. A temporary telegraph office was set up in the unburned portion of the city, to send out word that Virginia City was in flames.[14]

The Babcock chemical extinguisher, which had moved towards the Catholic church, was suddenly cut off by a wall of flame and had to be abandoned by the men of Eagle Engine Company No. 3, who turned to watch their shiny new apparatus totally wrecked by the fire. Nearby, a train of wood cars was caught at the north end of the track of the railroad and also destroyed. The heat of the fire near the Virginia & Truckee depot was so intense that the wheels of the railroad cars were melted to the tracks. According to the *Territorial Enterprise*:

> A wild scene was presented when the fire was at its height. Viewed from the elevated ground to the westward of the city was a sea of flames, from which vast columns of inky smoke rose hundreds of feet into the air. On all sides was heard the crash of falling roofs and walls, and every few minutes tremendous explosions of black and giant powder, as buildings were blown up in various parts of the town. Some of these explosions were so heavy that they are said to have rattled crockery and glassware in the town of Dayton, five miles distant.[15]

Adding to the general confusion and noise of the disaster was the sound of men, women, and children, some crying fearfully, dragging their salvaged belongings in a wild dash to safety. No sooner did they find a safe place to rest than they had to pick up and move on again as the flames advanced on them. Some had to move six to eight times, and with each move some important piece of property or cherished memento was left behind, giving them a smaller pile of belongings at each next stop. "Furniture of all kinds was consumed after it had been carried into the streets," the *Territorial Enterprise* reported, "and not a few pianos were thus abandoned to their fate after they had been carried out of buildings." Wells Drury recalled many years later that the fire was especially disastrous for him: "I lost everything I owned, including a trunkful of personal keepsakes, and my special pride, a pair of new gum-boots."[16]

Despite staggering setbacks, the Comstock firemen continued their battle, some losing several hundred feet of hose before finally being able to move into a position from which an effective stream could be played on the fire. As Dan De Quille described it:

> Water thrown into the midst of the flames produced no effect unless, as many thought, it added to their fury and fierceness. Although the firemen were at work with both hand engines and steamers, while yet but few buildings were involved, the water they threw upon the burning buildings might as well have been as much oil for any effect it had in checking the flames. The firemen were driven back from every point where they attempted to make a stand, and it soon became evident that no efforts of theirs could check the progress of the fire. It was such a fire as those which swept Chicago and Boston—a fire as fierce and uncontrollable as though belched up from the bottomless pits of the lower regions.[17]

The church bells and mine whistles continued to blast until they were silenced by the flames. The roof of St. Mary's in the Mountains took fire and sent flaming shingles flying about, setting fire to nearby buildings, until a charge of Giant Powder blew the roof off the church. According to a story by Charles C. Goodwin, an Irish woman rushed to the Consolidated Virginia Mine shaft, where John Mackay was working, and implored those there to save the church. Mackay is said to have told her, "Damn the church! We can build another if we can keep the fire from going down these shafts." Legend says he made good that promise with huge donations for the reconstruction and refurnishing of the church, but the diary of pledges kept by Rev. Patrick Manogue after the fire does not reflect a single entry in the name of Mackay or his mining company, though it does list several large contributions, including $1,000 from the wife of foundry owner John McCone.[18]

Mackay saved the shaft, but lost the hoisting works. The men in the mines were evacuated at the instant the alarm sounded, so that not a man might be left in the dark tunnels to face the possible fate of death in a burning shaft. The burning of the Ophir Mine hoisting works set the shaft on fire for a time, but it was kept from spreading too far by the hand engine of Knickerbocker Engine Company No. 5. The engine had been stationed at the mine's reservoir, and worked through the night to stay the progress of the blaze. At the insistence of John Mackay, he and James Fair had installed iron doors about twenty feet from the surface in the Consolidated Virginia shaft. These were closed and covered with two feet of dirt and debris, which kept the fire from snaking down that shaft.[19]

By eleven a.m., most of the central part of the city had been reduced to ashes. By two p.m., except for spot fires, the flames of the greatest conflagration ever to sweep through the Comstock had either been extinguished or burned themselves out. The destruction was complete. The burned district extended from just beyond Taylor Street on the south almost to the cemetery in the northern portion of the city. It extended from Stewart Street on the west to below G Street on the east. An area about three-quarters of a mile in length and one-half mile in width had been laid to waste. On the south, the last building to burn was Marye's new brick building on the west side of C Street, while the Black Brothers Building at Taylor and C Street on the east was the last to burn in that area.[20]

Among the principal buildings consumed were the homes of John Mackay and Judge Richard Rising; Derby and Gearhart's stables; several lodging houses; William Mooney's livery stables; the county buildings; the Virginia Hotel; Fulton Market; Piper's Opera House and saloon; Miners' Union Hall; Kennedy and Mallon's grocery and produce store; the Methodist, Catholic, and Episcopal churches; the First Ward School, John Gillig & Company hardware store; M. M. Fredrick's jewelry store; the Columbia House; Roos Brothers' clothing emporium; the Magnolia, Delta, Washington and Assembly saloons; the offices of the *Virginia Evening Chronicle* and the *Territorial Enterprise*; Reim's saloon; the post office and most of the mail; the Philadelphia Brewery and Carson Brewery; the Virginia & Truckee Railroad depot and several surrounding sheds; Consolidated Virginia hoisting works and mill; the new California stamp mill; the Ophir hoisting works; Haynie's lumberyard; many fine

Approximate boundaries for Virginia City's Great Fire of October 26, 1875. Star shows where fire began.

residences and several blocks of small homes; as well as all of the Chinatown district. "The loss of the hoisting works and mills is alone a calamity from which the city will not recover for more than a year to come, but when it is considered that nearly every principal business house in town is also destroyed, some idea of the blow it is to the Comstock may be imagined," the *Gold Hill News* announced to a world eager for word from the disaster site.[21]

The destruction included a major portion of the Virginia Fire Department as well. The volunteers had lost the engine houses of Virginia Engine Company No. 1, Eagle Engine Company No. 3, Knickerbocker Engine Company No. 5, and Nevada Hook and Ladder Company No. 1. Also lost were 3,500 feet of leather and carbolized hose, the Babcock chemical extinguisher of No. 3, the hose carts of Knickerbocker Engine Company No. 5 and Monumental Engine Company No. 6, as well as suction pipe for the engines, six cisterns, and thousands of dollars worth of furnishings, decorations, and firemen's paraphernalia kept in the various engine houses, along with the records of the respective companies. Nevada Hook and Ladder Company No. 1 also reported the loss of all its hooks and ladders, leaving the company with nothing more than an empty hand-drawn ladder truck, useless without equipment. Though devastated, the tired firemen of the city continued to respond through the remainder of the day and night with the equipment that had been saved, as winds whipped flames and sparks from hiding places amid the rubble of the city.[22]

Rumors ran rampant throughout the city, and an excited crowd gathered that night. It seemed to city officials that a riot was imminent, so the National Guard companies were called out to patrol, and martial law was declared. In order to assist the authorities, all saloons and liquor stores were prohibited from selling liquor, which seemed to many of the proprietors to be a great hardship. The majority acquiesced, however, and did not attempt to sell the "oh be joyful."[23]

Meanwhile, the newspapers did their best to quell the many rumors that were circulating. One false report was that two men were shot while attempting to set fire to the First and Third Ward schoolhouses. Another false rumor was that five men were killed by falling walls, that three Chinese women were burned to death, and that several of the city's most prominent citizens had also met their deaths. One rumor proved true, however, that a man had been seen trying to set fire to one of the hoisting works during the height of the blaze. This was the only documented case of incendiarism during the Great Fire. The newspapers further stated that the county records, which had been placed in the vaults of the courthouse, were found to be safe after the vaults had cooled. Today these records bear scorch marks and smoke damage from that 1875 fire, but are still in good condition. They are stored in the new courthouse built atop the rubble and old safes where they were cached during the fire.[24]

The loss of the Virginia & Truckee depot and buildings necessitated making the Gold Hill depot the central receiving point for all freight for individuals and businesses, some of which no longer existed. While most of the belongings of Virginia City's citizens were destroyed, some found their way into the hands of "thrifty individuals." It was reported that Gold Hill Constable Humphrey Symons "made a number of these people disgorge themselves of the fruits of their toil after the fire." Among the items he confiscated were clothing, full trunks, furniture, and even a full set of surgical instruments. The scavengers who had picked leftovers from the carcass of the city were themselves picked clean, and then encouraged to leave town as soon as possible.[25] The *Gold Hill News* reported the forlorn state of the city:

> Virginia City presented a sight last night such as has been seen by but few on this coast, a sight that all pray never to witness again. The fairest portion of the town, that in which was located all the principal business houses and large hoisting works that were the very life-blood of the city, and main source of wealth on the coast, was one heap of blackened, burning and falling ruins. Very few of the walls of the buildings that passed through the fire remained uninjured, and the greater portion of the noblest edifices were but confused piles of brick and mortar. Early in the evening the house of Banner Brothers, on the corner of C and Taylor streets, that had been burning inside for several hours, burst into bright flames, and the greater portion of the brick walls fell to the ground with a loud crash. The wind was at the time blowing in gusts that frequently amounted to a gale, and terrific whirlwinds swept over the city with a fearful noise, carrying coal-oil cans, pieces of tin roofing, and great clouds of dust and ashes wherever they went. The danger to life on this account was so great that very few had the courage and hardihood to pass through the smouldering ruins, in the early part of the night, but hundreds could be seen on the outskirts looking with rueful countenances on all that remained of their happy homes of the morning. And those remains, what were they? Simply the ground that the insatiate element in all its mad fury could not destroy, nothing more; for out of the thousands who in a few short hours were rendered homeless, very few saved more than

> the clothes on their backs, and will have nothing but the charity of the benevolent to depend on in their hour of need.[26]

On the western edge of the city, hundreds sat among their belongings in a daze, some weeping, some muttering, many just silently staring downhill at the city that only a few hours before had been one of the largest and best known on the West Coast. On the northeastern edge of the city, a number of men continued to work the brakes of the hand engine of Knickerbocker No. 5, in an effort to quell the fire in the Ophir shaft. They finally succeeded early the next morning. At three a.m., a rain shower assisted in cooling hot ashes and putting out many of the spot fires in fallen structures.

Within hours, relief efforts began. The first supplies to reach Virginia City from Carson City and Reno arrived the next morning in two freight wagons. Relief committee members were appointed and stations set up in the city's schoolhouses. Provisions, blankets, clothing, building materials, and other items were rushed from San Francisco, Sacramento, Seattle, Reno, Carson City, and a hundred other towns and communities. Some of the donors had previously been the beneficiaries of Comstock generosity when disaster had befallen them. As Eliot Lord wrote:

> Such a calamity is most deplorable, yet not without a bright side. The sympathy shown to the people of the city and the marvelous energy with which the ruin was repaired are memorable. The towns of Nevada and California contributed as with one impulse to relieve the distress of the sufferers, and as soon as supplies could be forwarded none were allowed to want for food, shelter and clothing; nor was their heartfelt liberality undeserved, for none was willing to impose upon the charity of their neighbors.[27]

Lord described the morning following the fire, when smoking timbers and debris littered the town, cooled only by judicious application of buckets or streams of water supplied from hydrants. Lumber that arrived on the railroad was placed on the black, reeking ground, and the work of building a new city then went on continuously all day and late into the night, in storms and fair weather alike. During the week following the fire, a tremendous windstorm blew down a number of the newly erected houses and stores, but the wrecks were cleared away as soon as the winds subsided, and building started anew. Sixty days after the fire, the principal streets running through the burned district were lined with new business houses, the majority of which were of a better design and construction than those destroyed, and habitable dwellings covered the intervening blocks.[28]

The night of the fire, Ferdinand Beck and his wife were forced to find refuge in an abandoned mine shaft. "It wasn't a comfortable night quarters," he wrote, "but we were safe out of the rain." The next day they were able to secure two rooms in a dwelling that had not been burned for twenty dollars a month, "and glad to have found new quarters."

Abigail Gardner, then a young girl on the Comstock, wrote to a cousin in New England:

> Mother and I returned to poor Virginia this morning from Reno, where we were staying with Mr. and Mrs. Pruitt. Father met us at the Virginia & Truckee Station. "Station" is an odd word to use. All the railroad's buildings were destroyed by the fire and now the agent's office is in a temporary wooden shanty while freight, etc., is stored in the open, its only protection a floor of timbers laid over the earth. Work has already commenced on the new station.
>
> We walked up Taylor Street to the Sprague's house on Summit Street. It was a fearful climb, but the few personal possessions we were able to save from the fire made such a pitifully small bundle that Father was able to carry all of it under one arm. The new clothing we ordered in Reno will follow us; soon, I hope. No hacks or porters to be hired, of course. All the men of the town who are not at the mines are busy tearing down walls, combing debris, and working on the new buildings which already are starting to rise.
>
> Mrs. Sprague received us kindly and led us to the room she had prepared for us, although the poor little Sprague children will have to sleep in the kitchen until we find more permanent quarters. This part of the city escaped the flames, and from here we can look down over all the parts of the town that were destroyed. Mrs. Sprague told us that as late as a week ago smoke still rose from smoldering ruins.
>
> After seeing to our settling Father left us to return to the mine. Later today Mother and I walked out to inspect what was left of our house on B Street, but the sight revived such poignant memories we soon left there. Mother found a fragment of a dish that had been part of the set given to her at her wedding and put it in her purse.
>
> We walked through the streets that once we had known so well but now they present a strange and confused appearance. Here a pile of charred wood, there a pile of new lumber, teamsters struggling their laden wagons through the maze. Everywhere there is a sound of hammering and sawing and shouting workmen.
>
> Many of the stores have reopened for business, a few of them housed in tents. Some of the storekeepers recognized us and all

> were full of optimism, saying that their new stores would be much better than the old, etc., so that Mother and I both were lifted a little out of our doleful mood. The sadness returned when we saw what was left of St. Paul's and St. Mary's Churches which had been blown up in an attempt to halt the flames, but then we met Father Manogue writing things down in a notebook and he said that already he had been pledged enough funds to build a much larger and more beautiful St. Mary's on the foundations of the old.[29]

The optimism of the merchants and townspeople was shared by many. Beck recalled, "It did not take long to rebuild everything within less than 1½ years, so many free volunteer helpers came from California, being it so near to winter. They estimated the loss of $10,000,000. Now I had experience in 2 fires. The one in Chicago and now in Virginia City." The difference between the two, he wrote, was that in the wake of the Comstock disaster he had not lost one day's work, due to the optimism and energetic efforts of those rebuilding, compared to the doldrums in the aftermath of Chicago's great fire.[30]

Those with large mining interests were publicly optimistic about the rebirth of the city, probably more in hopes of minimizing stock losses following the fire than in any expectation of a rich, renewed mining activity. John Mackay assured the world, "I have been through all the mines this morning and they are all right. There is no gas or fire in any mine connecting with the Gould & Curry." The word from the other mines was equally encouraging as work to rebuild got underway. Within a few days, the *Territorial Enterprise* concluded, "Amid what looks as if it ought to be enough to cause universal despair, there seems to be a brave confidence and unflinching determination to overcome the present misfortune."

A sour note in the chorus of optimism, however, was sounded by the insurance adjusters. An office was set up in which all claims could be handled. Yet, although claims were processed, none of the adjusters assured their clients that they would be paid the full amount of their coverage for the losses they suffered. In many cases, claimants were forced to sue their insurance companies to obtain payment.[31]

Despite this setback, Virginia City's residents were rebuilding their town. A new breath of life was coming to the Comstock, but at least one institution was doomed as a result of the fire and the resulting shift in attitude among the press and citizens of the city. No longer were the firemen of the city heroes in red shirts and leather helmets. The days of the volunteer fireman in Virginia City were numbered.

XVII. Bitterness and a Paid Department

The attack was almost simultaneous, as the two Virginia City newspapers unleashed a series of vitriolic editorials accusing the fire department of misdeeds during the Great Fire. They clamored for a paid fire department to replace the old volunteer system, which they claimed was responsible for the fire's destructiveness.

The *Virginia Evening Chronicle*, in one of its most venomous attacks on the volunteer fire department, urged the disbanding of the volunteer companies in favor of a full-time fire department, "the purpose of which will be to put out fires and not parade on holidays." The *Chronicle* complained, incorrectly, that "during the fire less than one half of the firemen were on duty, and even many of these occupied themselves in picking up inconsiderable trifles instead of obeying orders."[1]

Chief Engineer Frank McNair was made a scapegoat and blamed for the destructiveness of the fire. The local newspapers claimed that he had failed to use his water resources to best advantage, that he had failed to position equipment in strategic locations, and that he had failed to prevent fights among the companies, which supposedly led to the destruction of whole blocks while firemen battled each other.[2]

The charges were not true, but only the *Gold Hill News* took up the cause of the fire department. The paper complained, "In former times the firemen used to get at least a measure of credit, but now their only reward is slander and abuse." Editor Alf Doten's opinions were discounted by the critics, however, because of his long-time membership in Washoe Engine Company No. 4.[3]

As early as November 9, 1875, members of the fire department were speaking of disbanding and leaving the fire fighting to those most critical of the volunteer department and its chief. Eagle Engine Company No. 3 took the lead at a special meeting. Thomas Alchorn told the company that he was in favor of disbanding because the fire department's sacrifices and efforts to preserve property had been met with contempt and abuse. According to Alchorn, the *Virginia Evening Chronicle*

> . . . seemed only able to look at the actions of the Department from a prejudicial and unfair standpoint. The fact stood out prominently for anyone gifted with ordinary intelligence, that each and all of the late fires, over the mismanagement of which it cried so loudly, was due not to the inefficiency of the department, but to the inadequate supply of water, a state of things wholly the fault of the city.

Alchorn further suggested that with the gale which had been blowing at the time of the Great Fire, "all the water in the Truckee River would not have extinguished it in a very few minutes after it was started." Alchorn told the company that he was not one who believed in forcing his services on the community, especially when they were gratuitous, with the only reward being the satisfaction of doing good. The company as a whole agreed and immediately voted to disband, the first company to leave the old Virginia Fire Department.[4]

The *Gold Hill News* strongly supported Eagle Engine Company:

> We feel like saying a parting word in behalf of the members of this company and the Fire Department, who, in many a hard won victory over the raging fiery element, have proved themselves good men and true.
>
> There are men in this company, as in many others in the city, who have received injuries while struggling to save the lives and property of the citizens of Virginia, the scars of which they will carry to their graves; the only lasting testimonial of the regard of those they so unselfishly strove to serve. There are men in the company, to whom the tap of the fire bell, or the shrill shriek of the whistle, was, and is, an imperative summons to forsake a weary couch after a day of fatiguing labor, and face danger and perhaps death; and to their undying credit be it said they never shirked what they chose to make their duty. There are men in this company, as in the whole department, who never tire of exerting themselves in the public service, without asking a cent their whole demand being that their labors might be accepted and acknowledged in somewhat like the spirit in which they were generously given; and what to-day is their thanks? A press whose sense is if

> possible less than its decency, comes out and accuses these men of being petty thieves; neglecting their duty to pick up and appropriate the property they ought to be engaged in saving.
>
> Is it any wonder that, their property in which they took a pride being destroyed by the late fire, this base ingratitude has at last broken a spirit that was proof against hardship and danger, and the firemen are resolving to be no longer butt for the calumny of every inconsiderate and malicious idiot who chooses to turn his abuse upon them? Is it any wonder, we ask, that they have finally determined to see the city burn; and be damaged as much as it can rather than put up with such base ingratitude?[5]

By December, only 268 men had bothered to answer the roll call of the respective companies, out of more than 500 firemen prior to the fire. Compared to its previous state of readiness, the fire department had been left disabled. Nevada Hook and Ladder Company No. 1 had only its truck, without ladders, hooks, buckets, or other tools, all having been destroyed in the fire. The company's hall was in ruins and offered no shelter for what equipment had been salvaged. Obviously, the Hooks would be of little use during a fire.

Virginia Engine Company No. 1 had only 300 feet of very poor leather hose left, along with its hose carriage and Button & Blake hand engine, which had worked so hard during the fire that it was in dire need of repair. Without several hundred feet of good hose and at least cursory attention to the hand pumper, No. 1 would also be of little service during a fire. Young America Engine Company No. 2 was fit for service and ready to respond with its third-class Silsby steam fire engine, two hose carriages, 900 feet of hose, and equipment for 80 firemen. Eagle Engine Company No. 3, however, had lost its Babcock chemical extinguisher in the fire and had only 300 feet of unserviceable leather hose and its old Jeffers hand pumper.

The steam fire engine of Washoe Engine Company No. 4, which had failed to muster for the Great Fire, was still on the Divide for repairs and could not respond to an alarm. The company's hand engine had been sold to the Reno Fire Department, and the 1,000 feet of carbolized hose on its hose carriage would be of use only if the steamer were returned to service. Knickerbocker Engine Company No. 5 was in an equally difficult situation, without any hose for its hand engine or its steam fire engine, which had lost all but 12 feet of hard-suction hose. Monumental Engine Company No. 6, on the Divide, had lost a hose cart, but was still ready to respond to fires with its Button & Blake hand engine, Clapp &

Jones steamer, a hose cart with 1,000 feet of carbolized hose, and equipment for 80 firemen. In general though, the fire department was demoralized and virtually out of business.[6]

The city officials refused to pay for replacement equipment or, in some cases, even for repairs to existing equipment. With the Virginia Fire Department in such a condition, it is a miracle that the remainder of Virginia City was not wiped off the face of the earth in the eighteen months following the Great Fire of 1875. During that time, the merits of a paid fire department were debated in the press, in saloons, in the legislature, and in the surviving firehouses.

Even faced with the imminent disbanding of the old volunteer fire department, those companies which had not yet fallen out spit back in the face of their critics by reelecting Frank McNair as chief engineer of the Virginia Fire Department. The *Territorial Enterprise* called McNair's election" an outrage upon the property owners of this city," and declared the election to have been prompted only by a spirit of vindictiveness and defiance to the wishes of the public. The *Enterprise* continued its attack on the volunteer department, and especially Chief McNair, in the same vein:

> Toward McNair as a man we bear no malice, but on more than one occasion he has shown his utter incompetency as fire chief, and after such exhibitions to re-elect him is a direct insult to our people. And the matter goes still further. The firemen have no right to assess the property owners here thousands of dollars just to keep a pet in office. Insurance on fire proof buildings in this city now costs four percent. One and a half percent of this is the extra rate insurance agents demand because Frank McNair is at the head of the fire department, and further, in many instances, those agents have given property owners notice that their policies would speedily be withdrawn unless the Fire Department was at once reorganized. To this notice the firemen respond by electing a man chief, who by his misfortunes if not his faults, has forfeited nine-tenths of the property owners of the town. In the name of those property owners we call upon Mr. McNair to resign. His election has proved that the firemen like him as a man; now we ask him to show that he has the best interest of this people at heart by giving up his office. He must know that as a leader of the Fire Department he is not trusted by the men whose property he is in part custodian for. His own self-respect ought to suggest to him the propriety of resigning. If he thinks that past experience will enable him in the future to guard against former blunders, he is mistaken. Fire chiefs

> are not made to order; they are like orators, poets and painters—they have to be born fire chiefs. Mr. McNair was not, and the fact is costing the people thousands of dollars in extra insurance, even if there were no mistakes made. We hope Mr. McNair and the firemen generally will accept this in the spirit in which it is written. There should be no ill-feeling between the citizens and the firemen, but when the citizens grant every reasonable request of the firemen, they, in return, should respect the reasonable desires of the citizens.[7]

But McNair would not yield his office and the fire department stood squarely behind him. The *Territorial Enterprise* renewed its attack three days later by declaring, "With McNair as Chief Engineer, the Pacific Ocean would be useless." Maintaining that the city had suffered enough from the "pet" of the fire department holding office, and that "such a thing can no longer be endured," the newspaper once again called for the disbanding of the volunteers, creation of a paid fire department, and the resignation of Frank McNair.[8]

Undaunted, McNair set out to rebuild and improve the city's fire defenses. Despite political opposition to his holding the office, the chief engineer ordered new hose, and through contacts with a few influential persons around town, McNair was able to gain approval for construction of a system of hose houses, additional fire hydrants, and a new water system to supply them. McNair worked as if to vindicate his election as chief engineer. At his request, the city ordered a number of two-wheel jumper-type hose carts, to be strategically located around the town. It was at this time that the order was placed for two horse-drawn hose carriages from the Kimball Manufacturing Company in San Francisco, the first horse-drawn fire apparatus to be used on the Comstock. McNair also saw to the refitting of Nevada Hook and Ladder Company No. 1 with ladders, hooks, and other equipment, to be housed in the new engine house under construction. In this effort he profited from the influence and lobbying efforts of opera house owner John Piper, who also happened to be a member of the company.

It appeared that the Virginia Fire Department was on the rise once more, but the 1876 Fourth of July festivities and parade hinted at a different future for the volunteer fire department. None of the companies turned out a strong force of members, as they had in the past, although it was argued some were in San Francisco and attending the Philadelphia exhibition. The *Gold Hill News* recorded the procession:

> First came Hook and Ladder No. 1, or what the October fire left of the apparatus. The machine was gaily decorated, two ladders making a gorgeous triangle of red, white and blue over the truck. Eight or ten boys in white shirts and straw hats walked on either side, and the few remaining members of the old company marched in front in the orthodox red shirt. Following was the old hand engine, Virginia, drawn by six horses. On the old machine sat three little girls in white under a silk canopy of red, white and blue. An inscription in gilt on the rear end of the engine announced that the Virginia Company had been organized March 2, 1861. Washoe No. 4 followed, drawn by six horses. The engine was polished and decorated beautifully, and the men of the company in their red and black appeared to advantage. Knickerbocker No. 5, drawn by eight horses and in a blaze of polish and a mass of flags came next. The company turned out in full force. Following the Knickerbocker was Yellow Jacket No. 2 of Gold Hill, looking as fine as any with her six horses, small forest of flags and 60 men. Monumental No. 6, with eight horses, 40 men at the engine and 30 at the hose carriage, on which were seated two little girls and a boy in uniform, made a bright feature of the procession. The new hose company of Virginia, "Rescue Independent," brought up the rear of the Department, the members cool and comfortable in white shirts and straw hats and the machine very tastefully decorated.

Missing from the procession were the members and equipment of Rooster Hose Company No. 1, Young America Engine Company No. 2, Good Will Hose Company No. 2, Eagle Engine Company No. 3, Lincoln Hose Company No. 3, and Neptune Hose Company No. 5. The erosion of the fire department was becoming increasingly evident.[9]

Political forces were also eroding the old volunteer system. In Carson City, lobbyists in the halls of the Nevada Legislature urged passage of a bill disbanding the volunteers and authorizing the creation of a paid fire department for Virginia City. As soon as the bill cleared the legislature, the governor penned his name on it—signing the death warrant for the volunteer companies of the Virginia Fire Department. One by one, the volunteers began to dismantle the companies and sell the equipment that had helped protect the city from fire for so many years.

The city purchased the hook and ladder truck for $450. Young America Engine Company No. 2 offered to remain in service until the paid department could be organized, but demanded that its steward's pay be increased, or else the company would refuse to respond to any fire in the city. Knickerbocker Engine Company No. 5 made the same offer without the ultimatum. Instead, No. 5 requested the city purchase new hose for its steamer and hand engine, which the city agreed to do. Monumental Engine Company No. 6, on the Divide, was allowed to remain in service and given a $300 monthly subsidy, provided that the Monumentals would respond to fires with their steam engine in either Gold Hill or Virginia City.[10]

The reorganization of the fire department called for a paid force of not more than sixteen firemen, a chief, and two assistants. They were to be governed by a board of fire commissioners, which was to meet on a monthly basis. In February of 1877, Young America Engine Company No. 2 announced that it would formally disband in April. This left only Knickerbocker Engine Company No. 5 and Monumental Engine Company No. 6 on standby while the new paid department was organized and new equipment ordered and delivered. Eventually, Knickerbocker Engine Company offered the city its steamer, hand pumper, and other equipment for the new fire department. The city paid $8,500 for the apparatus, and also rented a portion of No. 5's new firehouse for temporary city offices.[11]

The legislation creating the Virginia Paid Fire Department authorized salaries of $200 per month for chief, $150 for the assistant engineers, and $125 for regular firemen. These amounts were two or three times the salary paid to firemen in eastern cities at the time, and an increase over what the volunteer fire chief had been receiving. In addition, city policemen were to receive an additional $25 per month, in exchange for responding to fires in the city when needed. The chief engineer also had the power to employ additional firemen and watchmen when needed.[12]

The selection of a chief remained to be made, however. The old prejudices against volunteer Chief Frank McNair surfaced when the town's insurance agents prepared a petition backing former volunteer Chief James K. B. "Kettlebelly" Brown for the office. To many, the petition offered an easy solution to a sticky problem, but the election itself was no easy matter.[13]

It turned out to be somewhat of a comedy. Commissioner Gallen nominated McNair. Commissioner Pottle nominated former volunteer Chief William Pennison. Commissioner Kaneen nominated Brown. President McKay nominated Thomas H. Alchorn, a member of long service in Eagle Engine Company No. 3. Commissioner Comstock, feeling somewhat left out, looked around the room and nominated Captain Lord. Kaneen responded that the choice was a good one, as the chief should always have the Lord on his side. Captain Lord, however, declined to be either chief or Lord on the chief's side.

The nominations were closed and the voting began. For seven consecutive ballots, McNair received Gallen's vote, Pennison got Pottle's, Brown had Kaneen's and Comstock's, and Alchorn had the support of McKay. On the eighth ballot, McNair got 1 vote, Pennison 2, and Brown 2. On the ninth ballot, Pennison received 1 vote, Brown received 4, and McNair received none. Brown was then declared the new fire chief of the Virginia Paid Fire Department. He was formally introduced to the board by McNair, who was subsequently nominated for the position of first assistant chief. But the disfavor of the insurance agents and others in the community prevailed, and John Crossman was elected first assistant chief of the new fire department.[14]

During the next few weeks, business included organizing the fire department; taking inventory of the equipment; drafting rules and regulations; and laying out new fire districts, with corresponding fire alarm signals to insure quick response. There was also the matter of appointing the new paid firemen. Frank McNair was nominated for a position as driver of one of the new horse-drawn hose carriages, but Fire Commissioner Kaneen said the driver of a hose cart should be a good, live man, that it was a different business from driving twelve or sixteen horses attached to a heavy wagon. McNair, a teamster by trade, took exception to this and lashed out, "I am not a dead man, I want you to understand." The matter was immediately tabled until tempers had cooled. McNair was eventually selected to be one of the town's paid firemen and served the new department through 1884, as both fireman and cart driver.[15]

After fighting its first fire, the new fire department received strong commendation from the press which had argued so hard for its creation. The *Territorial Enterprise* stated, "The manner in which the little fire was managed last evening showed skill and care." The *Virginia Evening Chronicle* said it was proud of the efficiency of the new fire department, and that "the chief engineer was deserving of commendation for his coolness and ability." Even the *Gold Hill News* said that the new fire department was composed of "practical and energetic men," and then with little subtlety noted that most of the men in the Virginia Paid Fire Department were former members of the old volunteer fire department it had replaced.[16]

The city had purchased the old engine house of Virginia Engine Company No. 1, near the corner of Sutton and B Street, and several

As part of the buildup of the new Virginia Paid Fire Department after the Great Fire of 1875, hose houses were built around the town; each one held several hundred feet of hose, spanners, nozzles, ladders, and other tools. (Author's photo)

pieces of equipment were on station around the city, including two-wheel jumper-type hose carts. Larger quarters were needed to hold all of the new equipment, however, quarters that would be centrally located in the city. The fire commissioners made plans for construction of the Corporation Fire House, where the city's equipment could be stored. The men and horses of the fire department would be quartered there, enabling them to keep a watch over the city and be ready to respond to any section in the event of an alarm. By August of 1877, the plans had been prepared. The board was ready to act on a proposal to build the new engine house on the hillside adjacent to the magnificent home of Robert N. Graves (now known as the Castle), who was superintendent of the Empire Mine.

The *Territorial Enterprise* told its readers that the new firehouse offered a commanding view of all portions of Virginia City, as well as the surrounding canyons and Upper Gold Hill. The structure would be of two stories, the lower one used as a storeroom and a hoserack for storing and drying hose, and to store the fire engines belonging to the city. The second story was to be used partly as a hayloft and partly as sleeping apartments for the firemen in charge of the hose carriages. The equipment could be moved either north or south via a slight incline driveway from B Street. On March 21, the newspaper reported:

> Work upon the building was commenced this morning and four carpenters are now at work to be finished by October 1 by contract. Total cost $2,000. Of this amount, $400 will pay for labor, extras ordered will cost $150, supervision by the city clerk $100, chimney about $50, making a total of $700. The remaining $1,300 of the estimated cost will be for materials, all of which the city supplies, the building being solely for the use of the fire department.[17]

Except for installation of new gas light fixtures, the firehouse was ready for occupation by October 23, 1877, nearly two years to the day after the Great Fire. The equipment was housed without the fanfare that had attended similar occasions in the days of the volunteer department. The paid department settled into its new quarters and settled down to keeping the city safe from the "fire fiend."[18]

Mayor C. H. Belknap reported to the aldermen that, with the installation of the new water system, completion of the new Corporation Fire House (which finally cost $2,930), and purchase of horse-drawn hose carriages and other improvements, "the superior facilities possessed by the city for the extinguishment of fires by means of its powerful water

The Cork Corporation House, home of the new Virgina Paid Fire Department; it featured a watchtower, catwalk, stables, and housing for firemen as well as for two, four-wheel, horse-drawn hose carriages, the old hook and ladder truck, No. 5's old steamer, and a hose sleigh. The firehouse was torn down in 1948, suffering the indignity of being called a fire danger. (Greenhalgh family)

works has rendered the use of engines unnecessary." The old Amoskeag steamer that the city had purchased from Knickerbocker Engine Company No. 5 was placed in virtual retirement in the back of the firehouse. Along with the hook and ladder truck, it seemed more a relic than a working piece of fire apparatus.[19]

The bell from the B Street house of Virginia Engine Company No. 1, purchased by the city, was removed and placed atop the Corporation Fire House. The B Street station was remodeled to permit the housing of two horses and a carriage with a special set of suspended, hinged harness, for quick hook-up of the team in the event of an alarm.

The new Virginia Paid Fire Department seemed well on its way to efficient operation. Yet, although the volunteer department had passed from service, the volunteers had not passed from view. As the volunteer fire companies began to disband, their members joined to form the Virginia Exempt Firemen's Association. Membership was restricted to only 250 members at a time, all of whom had to have seen at least three years' active service in the old fire department.[20]

The Exempts elected William Pennison their first president, then purchased the old engine house of Knickerbocker Engine Company No. 5, on North C Street, to use as a meeting hall and museum. They began amassing a collection of relics from the companies of the old volunteer fire department. Virginia Engine Company No. 1 donated its hand pumper, the first fire engine in Nevada. Washoe Engine Company No. 4 donated furniture, helmets, belts, and other mementoes. Nevada Hook and Ladder Company No. 1 brought a silver trumpet, awarded many years earlier to one of its members for selling the most tickets to a fire department benefit. J. W. Hemenway, a member of Eagle Engine Company No. 3 since 1863, ordered an engraved silver trumpet to present to the Exempts to commemorate the founding of the organization. Other companies and individuals also donated to the museum, with items coming not only from the old fire companies of Virginia City, but also from those of Gold Hill and from other companies throughout the United States. All of their gifts were arranged about the hall of the Exempts, for public inspection.[21]

In addition, the organization took possession of the fireman's cemetery north of the city and continued to care for the ground, which the volunteer fire department had purchased in 1868 for $50. The Exempts created a fund for members who were down on their luck, or seriously ill and out of work. A visiting committee, responsible for visiting members who were ill or hospitalized, or the families of those who died, was appointed each year. The Exempts also buried their members with pomp and circumstance, and continued many of the traditions of the old volunteer organization, including turning out for parades and special occasions, and playing host to visiting firemen from other cities.[22]

Those Exempt members formerly belonging to Eagle Engine Company No. 3 took responsibility for flying the flag from the flagpole atop Mount Davidson. This pole had been erected during the Civil War years, at the direction of Chief Tom Peasley, who had also been a member of No. 3. On each holiday, members of the old company would climb the steep mountain before dawn, to hoist the colors and fire off a salute of Giant Powder.[23] In 1878, a committee of Exempts comprised mostly of

Frank Conlan and Dave Tweedie on City Hose Cart No. 1, which was built in 1877 by the Kimball Manufacturing Company in San Francisco. Photo taken in front of the Storey County Courthouse in 1908. The carriage is now on display in the Comstock Firemen's Museum. (Comstock Firemen's Museum)

members of the old Eagle Engine Company erected a new flagpole to replace the earlier one which had snapped in a windstorm. The committee made its way to the top of Mount Davidson before daylight on July 1, 1878. They were soon followed by more than 300 persons from Virginia City, Gold Hill, and Carson City. The first ladies to arrive at the site at 6:30 a.m. were Katie L. Emerson of Carson City, and Libbie Beebe of Virginia City. The new 85-foot flagpole, topped with a silver glass ball, was made of iron pipe covering a wooden core. The pipe was donated by Adolph Sutro, white paint was donated by Frank Parke, and wire cable for hoisting the flag was donated by Col. Gillette of the Savage Mine.

The flagpole was placed into a hole that had been blasted into solid bedrock at the peak of the mountain. At the bottom of the hole was a sole plate of iron, upon which the new pole was to rest. Once in place, the flagpole was set up with guy ropes and the hole was filled with masonry set in cement. When this was done, Miss Beebe advanced to the base of the pole and broke a bottle of wine over the iron to christen it. Miss Emerson, assisted by Thomas Peasley's brother, Andrew, raised the flag in a bundle to the top of the mast. When the word was given, the lashings were pulled from the ground, and the flag, which had been donated by James G. Fair, unfurled gracefully in the breeze. A signal cannon was fired at the same time and the echo could be heard at the base of the mountain in Virginia City. The crowd signalled its approval with three hearty cheers. A gun was fired for each star in the flag, and each volley was answered with a cheer. The flag flew for about two hours, while the crowd attacked a sumptuous banquet that included beer and other beverages donated by Richard Brown, John Deininger, and the firm of Gumbert & Webber, and food donated by Spiro Vucovich and Charley Mueller. Today, the tradition of raising the flag on holidays is carried on by members of Virginia City's volunteer fire department, who rise before sunlight and make the trek to the top of the mountain to raise "Old Glory."[24]

The new paid fire department weathered its first years with relatively few problems or major fires. Despite the efficiency of the paid men, though, there were still fires to be fought by the volunteers, and these provided many anxious moments for the citizens of Virginia City. While conflagrations did not seem to occur as often, or to be as destructive as in years past, there were still fires that caused considerable damage and resulted in controversy.

On March 13, 1883, Piper's Opera House exploded in flames following a masquerade ball that had lasted until the wee hours of the morning. John Piper put out the gas lights at three a.m., and retired to his apartment in the opera house. At five a.m. "Old Dave," a retired sailor employed at the Delta Saloon, was headed to his home on A Street when he noticed flames curling about inside the opera house and raised the alarm. Immediately, the whistles of the various hoisting works and the fire bells alerted the town and the paid fire department.

Despite the fact the alarm had been given promptly, the firemen were slow getting to the fireground. Once in position, they put six streams to work combatting the blaze, but the fire had gained considerable headway. In a very few minutes, it engulfed more than three-quarters of the building. John Robertson, watchman for the Nevada and California banks, rushed inside the burning structure and stumbled around in the smoke until he managed to arouse John Piper, who was still asleep and unaware of the peril he faced. Both made a hasty exodus as the flames advanced upon them.

The fire was spectacular, lighting up the early morning sky and sending an inky column of dense smoke and burning embers straight up into the air. Hundreds of spectators rushed to the scene, but those living in or owning adjacent buildings were too busy gathering their belongings to watch the progress of the fire. The memory of the 1875 calamity was still fresh in their minds. Several lengths of hose burned and burst as the firemen made their attack. The thick smoke curling out from the front of adjacent buildings hindered fire-fighting operations, but they continued to pour solid streams into the midst of the inferno, turning water upon adjacent buildings to keep them cool and prevent the spread of the fire in any direction. In so doing, the firemen wrecked the saloon connected to the opera house, which was also owned by Piper. They flooded the surrounding buildings with water and broke every front window as they directed the heavy streams into them.

The firemen were fortunate that there was no wind to drive the fire into other buildings, but their efforts to save the opera house failed. Within a short time, the opera house was in ruins, with losses estimated at $35,000. A hose and firemen were kept at the scene all the next day, to cool hot spots and prevent a rekindling of the fire. Piper had no place to stay, but the community rallied to his support, offering him a hotel room, a house, and a benefit. Instead, Piper continued his business dealings. He arranged for the use of Cooper's Hall for the act scheduled to perform the next day, and immediately began plans for the reconstruction of his opera house.[25]

While the paid department fought fires, it also battled the whims of politics. Chief engineers came and went as the political winds shifted. Chief James K. B. "Kettlebelly" Brown was replaced by John Rush in 1880. Rush, in turn, was replaced a short time later by William Pennison. John Riorden succeeded Pennison, and promptly marched into controversy in 1883, with his first major fire. A blaze near the Gould & Curry Mine destroyed five frame buildings before it was halted, bringing the chief adverse attention from the press and the town. The inefficiency of the fire department was not to blame, however. As Riorden apologetically (and ungrammatically) explained to the county commissioners, the firemen were victims of circumstance:

> At 7:30 a.m. an alarm was given at the Corporation House of a fire near the Gould & Curry below the railroad. It proved to be in a

small house in the rear of John McCausland's near D St., the house being very low and in a hollow with large buildings in front. The man on watch was deceived and located it too far east. In consequence, the [hose] carriage was ran rapidly down Taylor to D St. When the fire was seen to be west of D St. the men drove rapidly for the Curry hydrant, that being the only one on D St. in that vicinity. On our arrival we found it engaged by the Curry men. We then attempted to get up Flowery on to C St. The hill being steep the horses was unable to dray it up. We where compelled to return back D to Taylor, up to A St., then on to Flowery thereby loosing much valuable time.

We took the hydrant on the corner and also the one at Cole's Drug Store, ran the first line of hose down between McCausland's and the church, turned on water when the hose bursted. I stopped this with one of Hayse patent clamps, when the second line also bursted and was secured in the same manner but bursting again was uncoupled and thrown asside. By this time the fire had made rapid headway. Dense clouds of smoke and flame came across C St. making it impossible to remain in this position. We took one line north, the other south with a view to confining the fire in its present limmits, Viz: Zigler's on the north and Chinese wash house on the south.

I sent the assistant engineer on D St. to get water their to prevent the fire extending east. He reported the Curry on the northeast and the Savage on the southeast doing well. I then gave him charge of the south end of the fire with instructions to hold it from spreading further south. He drove to Silver St., draged hose down and put them on the double hydrant. At this time No. 6 [Monumental Engine Company No. 6] volunteers got out a line in front of Savage works. With those four lines, he efectively stopped the fire at the south.

I resolved to stop the fire at Zigler's brick building on the north. There are several frame buildings in the rear of this which I feared might carry the fire around me, but those where kept constantly wet by the Curry stream. The fire was now so hot that the buildings on the west side of C St. where continualy taking fire, compelling me to keep alternating between them and the main fire. I ordered all the available hose on B St. and got out two lines in the rear of those buildings and thereby saved them. The fire raged some time when the walls of the Virginia House fell. When we closed in on the fire with all streams, and shortly had it under control and finly out, all those inflammable wooden buildings where confined in a space of 120 by 60 feet. The losses are (as per assessor's roll): $5,000.

Chief Riorden concluded with high praise for the men of his department, the volunteers from the Savage and Curry works, the members of Monumental Engine Company No. 6, Chief John Marks, Gold Hill Fire Department, and the many citizens who assisted in suppressing the fire.[26]

The loss of five buildings was too much for the town, no matter what the circumstances, and the commissioners were once again pressured to change fire chiefs. Riorden held his office for only three months in 1883, being succeeded by George W. Hanbridge, whose term was also three months. Finally, William Pennison, who had held the position of paid chief before, and who had served as chief of the old volunteer fire department, was selected to be chief of the Virginia Paid Fire Department. Pennison, who had been born in London, England, in 1836, was a stabilizing influence on the department and retained the chief's job for more than thirteen years.[27] Pennison's term, too, was to include some anxious moments, as the "fire fiend" was loosed on the town from time to time.

During this time of transition for Virginia City's fire-fighting forces, Gold Hill still relied upon its volunteers. The Gold Hill Fire Department remained active in 1883, despite the disbanding of Yellow Jacket Engine Company No. 2 the year before. Liberty Engine Company No. 1 continued to dominate the remainder of the fire department, continually electing its nominees for chief engineer.[28]

In 1885, the old Jackets hose carriage was taken from storage by Gold Hill Chief J. F. Gladding and put in the care of a new company, Divide Hose Company No. 2. The new company was to be an integral part of fire fighting in both Gold Hill and Virginia City throughout its career.[29] Before it even had hose, the new company responded to a blaze near the Fourth Ward School, in a tailor shop occupied by Gustave Friedlander. The *Territorial Enterprise* reported that the newly organized company gained the honor of putting "first water" on the fire by taking hose from a nearby hose house and attaching to the hydrant nearest the fire. The hose was too short to be effective, and only by stretching it to the maximum could the new volunteer company get even a spray on the burning structure. When the paid fire department arrived, its hose was connected to the Divide company's, and made short work of what could have been an extensive fire. With three streams on the blaze, damage was minimal.[30]

Each year both the paid fire department and the volunteers from the

Divide Hose Company No. 2 was organized in 1885 and was given the old hose carriage that had belonged to Yellow Jacket Engine Company No. 2. The Divide firemen were active until they disbanded in 1937. (Paula Hardy)

The fire in Rosenbaum's furniture store on C Street, July 26, 1888, caused dense clouds of black smoke to fill the streets. It was quickly put under control by the city hose carts and the volunteers from Divide Hose Company No. 2. (Nevada State Museum)

Divide and Gold Hill were called out for numerous similar fires. There were still spectacular blazes to contend with, however. These occasionally threatened whole blocks at a time, despite the best efforts of all the fire fighters.

Sometimes the fires were of such magnitude that the old volunteers came out of retirement to help the new paid department. One such occasion was the fire in Rosenbaum's Furniture Store on C Street, July 26, 1888. The blaze was started by children playing poker behind mattresses stored in a small closet in the back of the store. They used matches instead of chips, and one of them ignited when stepped on. The door to the storage closet was closed when the fire was discovered, thus cutting off the oxygen supply. Once the preheated smoke and gases were ignited, there was a backdraft explosion, which threw a sheet of flame and smoke out onto C Street.

The fire department arrived quickly. With the assistance of Divide Hose Company No. 2 and a crew from the Gould & Curry works, directed by James G. Rule and Hebe Holman, they soon had five streams on the fire. Three came from C Street and two from D Street. Ladders were placed on the porch of the one-story furniture store and streams directed wherever flames jumped out. Immense volumes of dense, pungent smoke rolled from the front of the building, reminding the firemen that mattresses of pulu, hair, wool, moss, and cotton were burning, together with rugs and other furnishings.

The fire on the C Street side had been virtually controlled within twenty minutes; but on the D Street side, it had moved into wooden buildings and threatened to eradicate a whole block. The blaze moved under porches and roofs, creating a problem for the firemen trying to get streams of water on the flames. They continued to maneuver their hoses to the best possible position, moving through Ritchie's barbershop and Young's saloon and trying to prevent the fire from igniting paint and varnish in the nearby shop of Samuel Porteous. At one point, the fire moved to the upper story of the rear portion of the Pacific Lodging House, setting fire to a number of mattresses there. A miner sleeping in the building was awakened in time to escape certain death, while George Hatch and Johnny Dunlop moved along the roof of the building until they could climb over and down, clinging to the eaves, to let themselves in through the end window of the upper story. Once inside, they quelled the blaze by throwing out the burning mattresses and stomping out the spot fires inside.

Despite the great volumes of smoke, damage was minimal, according to the *Territorial Enterprise*:

> Good work was done at all points, as the charred and blackened ruins testify. In the front room of the furniture store the walls, lower floor and ceiling remain in place, though damaged in places. The frame store in the rear had the roof burned off at the point where the fire originated, and the sides were burned out of it for a considerable distance, yet (owing to the flood of water thrown in) the floor was not wholly destroyed. The greatest damage in the Singleton lodging house (Pacific), however, was from water, and the breakage of furniture and crockery. Every part of the building was completely flooded.

During the height of the fire, when the Pacific seemed doomed, a man came down the stairs with a bundle of crockery, glassware, and other items tied up in a shirt. The load was heavy so he dragged it down the stairs, crashing on each step, and finally landing on the sidewalk with a noise "resembling the celebration of some big god's anniversary in China-town, and the crash as it struck the sidewalk resembled the concluding explosion in an artistically conducted fusilade of firecrackers." When everything was over, more than $15,000 of damages were counted, $11,500 at Rosenbaum's alone.

The *Territorial Enterprise* reporter, Alf Doten, was himself an old fireman. He delighted in telling readers that the fire had "brought out many of the ancient war-horses of the old Volunteer Fire Department." Doten's account continued:

> After the fire was pretty well subdued, C. C. Thomas and John Egan were found taking something stimulating. "Ah!" cried they, "no reporter came around to see us work, but one is sure to be on hand to bear witness that we took an antidote against being stricken down by terrible colds." Mr. Thomas needed an "antidote," for he was struck and knocked over by a powerful stream of water with which he was obliged to wrestle for some time before he could get away from it. Mr. Egan also looked as though he had been trying to climb a stream to the roof of some three-story building. Indeed, scores of persons received thorough wettings, and antidotes were in brisk demand.

It was also reported that the crew from the nearby *Territorial Enterprise* had turned out to help fight the fire, and that they, "got as wet as so many drowned rats and probably saved as many stoves, smoothing irons and fire shovels, and smashed as much crockery and glassware as any other crew." One of the newspaper's fire fighters, Billy Bartlett, escaped injury during the blaze when, walking across a roof in thick, black smoke, he fell twelve feet through a skylight to the floor below.[31]

The year 1890 was a particularly lively one for the firemen of the Comstock. The Fourth of July turned into one continual fire fight, as the firemen responded to five alarms. One of these was to the Lone Star Stables, owned by J. B. Lynds. It was located on the Divide, virtually on the doorstep of the Divide Hose Company house. One of the hose carriages of the paid department was being driven back to the Corporation House when the alarm was sounded. The carriage, completely manned and ready for duty despite its decorations, quickly headed towards the fire. At the same time, the department's second carriage left the Corporation House at a dead run. It was one of the liveliest races since the days of the volunteers. Divide Hose Company No. 2, however, was first to get water on the stables, which had been built in 1862.

Chief Pennison directed the attack. At one point he called for additional hose stored in the old B Street house of Virginia Engine Company No. 1, which was used as a reserve station by the paid fire department. Despite the best efforts of the Gold Hill and Virginia City firemen, however, the stable was completely engulfed in a matter of minutes. Several firemen attempted to save the frightened, neighing horses inside, but to no avail. All attempts were futile, as the firemen were driven back by thick smoke and a very hot fire fueled by one hundred tons of hay and several tons of barley and other feed. Only three horses managed to escape cremation, by leaping fences and dashing along C Street with manes and tails blazing.

During the height of the fire, a blazing flag dropped on the roof of a stable across the street belonging to the Commercial Soap Works. Within seconds that building, too, was a mass of flames. The firemen worked quickly to prevent the spread of the fire, despite the fact that the wooden buildings in that area were tightly crowded together, including the hall of Divide Hose Company No. 2. In all, nineteen horses, a cow, and a calf were burned alive in the fire, with losses totalling more than $14,750. Lynds, a member of Divide Hose Company No. 2, had escaped with only a trunk of clothing and a few books. Within a year, however, he was back in business, with a new stable on the Divide.[32]

The cause of another great 1890 alarm was a fire in the chimney of the French Rotisserie restaurant. It burned a large hole in the roof, and huge clouds of dense smoke filled the street, but it was quickly extinguished without much fanfare. On another occasion that year, however, images of past fires and problems were suddenly flashed before the citizens of Virginia City, when an incendiary was reported "abroad" in the town. It was feared, the newspapers wrote, that the city would again fall prey to the "devil in human shape" which had plagued the community during the 1870s. After several buildings were discovered to have been the target of the arsonist, stern warnings were passed along the streets and in the newspapers, backed up with more frequent police patrols. The attempts gradually became fewer, and finally stopped altogether.[33]

The frequency of fire in August of 1890 kept the firemen of the Comstock busy and provided spectacular, if somewhat destructive, entertainment for the people of Virginia City. Three fires broke out within three days of each other, causing those who manned the hose carts and carriages of the Virginia City and Gold Hill fire departments many anxious moments. One was caused by the owner of a paint shop behind the Odd Fellows' Hall, who left a lighted tobacco pipe in a shirt pocket after hanging it on a peg. The fire was discovered by a passing Chinaman, whose excitement and difficulty with the English language

An extremely rare action photo of a fire in Virginia City in the 1890s. It shows the hose carriage of Divide Hose Company No. 2 and firemen on the roofs and balcony of buildings on C Street. (Greenhalgh family)

caused a great pandemonium. When Special Officer Coryell finally entered the premises, after ascertaining the problem from the excited Chinaman, all he could find was a lot of smoke and the burning shirt, still on the peg in the center of the room.

The second fire alarm was sounded when smoke was seen billowing from one of the front rooms of the International Hotel, on C Street. The alarm proved unnecessary, however. Spectators gathering on the street below the "burning" room observed a rocking chair come flying through the window, blazing like a rocket takeoff. A lamp which had been left burning all night had ignited the chair and some other contents of the room, all of which were jettisoned by the owner.

The real excitement came with the discovery that Lonkey's wood yard, on the northeast corner of Mill and D streets, was afire. Even Divide Hose Company No. 2 made the run from the southern edge of the city to assist the city hose carts in fighting the blaze, which started in a shed and woodpiles on the east side of the yards. The *Virginia Evening Chronicle* reported that the companies quickly put three streams of water on the fire from the west side, thus confining the flames to the area of their origin. A large quantity of cordwood was charred on the top tiers, and minimal damage to the steam engine and boiler was reported, but the loss only amounted to $1,000. The newspaper reported that the scene "was densely thronged with people, which included a large delegation from Gold Hill." The fire was out, but Lonkey's was to suffer virtually the same losses two years later, when the Divide and city companies again responded to the sound of the alarm, in virtually the same location in the wood yard. This second time, Lonkey lost his steam engine, saw, and chopping machine as well.[34]

Chimney fires were a continual cause of alarm, but were generally extinguished quickly, with a minimum of damage. On August 16, 1892, however, a burning chimney at the corner of C and Silver streets proved nearly fatal, and certainly disastrous for two of the city's paid firemen. The alarm came in shortly before eight a.m. Emory Smith drove the two-wheel, horse-drawn hose cart out of the Corporation House, with Harry Nye on the rear step. About sixty feet down the driveway, a paper blowing in front of the horse frightened it so that it veered and pulled the cart over the embankment to the street below. The cart rolled over twice before coming to a stop. Nye was crushed under the weight of the cart and its load of hose.

Still alive, but in very serious condition, the luckless fireman was carried to his house on C Street near Carson, where a team of doctors

Two-wheel hose cart on B Street, circa 1888, near the driveway to the Corporation House. At right is the home of James Fair. (Nevada Historical Society)

attended him. In addition to a broken nose and two broken wrists, his right leg had been crushed and broken above the knee, with a bone protruding six inches. The physicians tried to save Nye's leg, but amputation was necessary. Smith's injuries were serious, but he was able to walk away from the accident and was off work for only seventeen days recuperating. The horse had bruised legs, but continued in service; the hose cart had suffered only minor damage.[35]

Throughout these final years of the nineteenth century, the old traditions were maintained by men still proud of their volunteer heritage. Annual dress firemen's balls were held in Miners' Union Hall in Gold Hill, to help raise funds for the purchase of new equipment. The Comstock firemen continued to meet with their old friends in Carson City, in the Warren and Curry engine companies. They attended formal dinners and balls in Carson City, or travelled to Bowers Mansion in Washoe Valley to participate in holiday games and picnics with their friendly rivals.

As late as 1890, when death came to a member of one of the companies, all of the firemen mourned in the traditional fashion. When Charles McDermott, a member of Divide Hose Company No. 2 was shot, more than one hundred uniformed firemen from the Gold Hill Fire Department marched to the cemetery and placed floral tributes on the grave of their comrade.[36] As in the old days, the firemen still enjoyed a good parade when they were not battling a blaze, and the appearance of the Comstock fire companies with their equipment was always a favorite with spectators. When the Liberty Hose Company and Divide Hose Company decorated their carriages for the Fourth of July parade in 1891, it prompted one Comstock resident to proclaim, "It's no use talking, a procession ain't a procession if the fireboys ain't in it."[37]

The tradition of rivalry between the companies was maintained as well, though for the most part they were friendly rivalries unless it happened to be time for department elections. Late into the century, the struggles between candidates and fire companies in Gold Hill were still severe. They were every bit as intense as the contests for offices in the old volunteer fire department of Virginia City, with considerable hustling for votes. The fire bells pealed as the firemen gathered for a celebration when the election results were announced. Until the next election, their disagreements and fights were forgotten. One particularly hard-fought election campaign for chief of the Gold Hill Fire Department was waged between Frank Fox and Henry Baglin. The former received 80 votes, and the latter 57 votes. That 1896 election caused the *Virginia Evening Chronicle* to comment:

> The election stirred up Gold Hill more than anything that has occurred there for a very long period. All day long, while the election was in progress, the town was full of life and excitement, and after the results became known, a celebration was inaugurated which lasted all night. Bonfires were lighted in the streets, pistols and guns were fired off and a general jollification was conducted by the adherents of the successful candidates. The fire boys who supported Fox worked hard for him and their victory followed upon the heels of a closely contested fight. They were consequently much pleased with the result of their efforts.[38]

Despite such jubilant celebrations, and the maintenance of the grand traditions of the old fire companies, the volunteer fire departments of the Comstock had passed their zenith. The fortunes of the Comstock had declined also, and with the slowdown of the economy came reductions that would drastically affect both the paid and volunteer fire departments. The position of assistant chief of the paid fire department was eliminated, and, with the consolidation of county government, the Board of Fire Commissioners was absorbed by the county commission. The pay for extra firemen for the paid department was reduced, as were the subsidies received by the volunteer fire companies. The Virginia Fire Department was reduced to a force of only six men, who also had responsibilities as policemen for Virginia City. The fire chief also became the chief of police, thus consolidating further and reducing the effective force of the fire department. Alice Byrne, whose husband John served as a relief driver and later fireman on the city hose carts, recalled that when there was trouble in town, the fire department was called out to support the police officers on duty. On many occasions, City Hose Cart No. 1 was requisitioned to assist in apprehending disorderly persons, who would be handcuffed to the back of the hose carriage and hauled off to the jail in the courthouse.[39]

As the Comstock moved into the twentieth century, activity in the mines slowed to a crawl, the population shrank, and there was no longer a need for a large fire department. By 1901, Chief Michael E. Nevin* reported that only he and a force of four firemen were on duty, with salaries for the entire force amounting to only $525. According to the reports filed by Nevin, the small department averaged only five fire alarms per month. Generally, two of those were false alarms.[40]

In order to augment the shrinking paid fire department, but not provide a full subsidy for volunteer fire companies, the Storey County Commission authorized payment of up to three dollars per person when volunteers worked during an alarm. Thus when the fire department was called out to a major blaze, such as that which claimed the International Hotel in 1914, there was plenty of help on hand for the paid fire department, in addition to the volunteers responding from Gold Hill and the Divide.

While the cause of the 1914 hotel blaze was never really determined, a Chinese cook recalled after the fire that the electric wires in the famous old hotel were continually "burning." When the alarm bell sounded at the Corporation Fire House, the sky over Virginia City was already filled with dense smoke and a bright orange color from the flames. When the city hose carts of the paid fire department arrived at the scene, Chief

---

*Michael E. Nevin served as chief of the Virginia Paid Fire Department from 1901 to 1906, and 1911 to 1912. In 1983, his great-grandson and namesake was chief of the Storey County Fire Department, and another great-grandson, Dennis Nevin, was a paid fireman in Virginia City.

Although the heat was so intense that it melted glass in windows across the street, frozen hydrants and hoses full of icy slush hampered the efforts of the paid and volunteer firemen to fight the 1914 fire which destroyed the International Hotel. (Alice Byrne)

Charley McQuigan found that all of the fire hydrants in the vicinity had been frozen in the winter air. Orders were given to lay hose and chop ice from the hydrants to try to get them working. Meanwhile, volunteers from Liberty Engine Company No. 1 in Gold Hill and Divide Hose Company No. 2 made their way along the icy streets to the blaze and began laying hose lines from a few hydrants found to be working. However, once they got into position, the Gold Hill volunteers found their hoses freezing. Only slush came from nozzle tips. Firemen had to pound the hoses with heavy mallets to keep the water flowing, despite the intense heat of the fire, which produced so much radiated heat that windows in buildings across the street from the large hotel were melted. Only through the terrific efforts of the firemen were the adjacent buildings saved from destruction.

Fortunately, there had been only about a dozen persons in the hotel when the fire broke out. They were warned by Charlie Young, who rang the hotel fire alarm gong and raced through the halls of the six-story structure. Chief McQuigan sent men into the 160-room hotel to help rescue guests and employees. Some made their way to safety or waited at windows until ladders could be raised. Hotel manager George King crawled through smoke-filled halls with a handkerchief over his face until he finally reached a window where a ladder had been put up. Guest Fred Davis thought an earthquake had struck the Comstock when he was awakened by the partial collapse of his floor of the hotel. When he opened his room door, he was confronted by a sheet of flames, which forced him to retreat through a nearby window, carrying his pants in his hands. Flames rushing up the elevator shaft created a roar that was described as being "like a railroad train."

Adjoining buildings were also imperiled. Radiated heat caused Piper's Opera House to catch fire several times, to be extinguished by one of the twelve streams of water directed at the main fire. Burning embers were carried aloft to land on nearby roofs, which were protected only by a light coating of fresh snow. When the fire reached its height, the C Street wall collapsed straight down into a massive pile of rubble, causing a great shower of sparks and embers. The paid firemen, Dave Tweedie, Frank Conlan, and John Kenney, worked hose lines and helped direct the volunteer efforts as other parts of the building came crashing down. The south wall fell into the middle of Union Street; and the north wall of the hotel collapsed through the roof of the adjoining Sexsmith Building, breaking through the second floor and virtually demolishing Wrenn's

Saloon on the ground floor. By this time a freezing rain began to fall on the battling firemen, who continued to pour water and slush into the debris until the danger of the fire spreading had been minimized. Within two hours, one of the most famous hotels in the West, and the toast of the Comstock in its day, had been reduced to a pile of rubble for at least the third time in its history. The next day Chief McQuigan ordered the remaining walls to be demolished as a potential hazard. Men with a hose were assigned to watch for potential outbreaks from hidden embers as cleanup of the wreck of the International Hotel began. Wagons hauled what was left of the once magnificent hotel to a nearby hillside and dumped the debris into a ravine.[41]

Continued reductions in manpower had forced the retirement of City Hose Cart No. 2 and its team of horses. The latter were put out to pasture and eventually sold, but the horses were hardly ready for their forced retirement. Old Gray, one of the large horses used on City Hose Cart No. 2, was an unusual horse, always enthusiastic about the run to a fire. Following his sale, Old Gray was put to work hauling a lumbering old furniture wagon. Three days after his forced retirement from the department, Old Gray was hitched to an express wagon piled high with furniture and other goods. The serenity of the scene was suddenly shattered by the sound of the fire alarm. Old Gray waited about the length of time that he thought it should take to be hitched to his old hose carriage, then bolted for the fire. The furniture on the wagon was scattered to the winds, from one end of the street to the other, but Old Gray reached the fire before City Hose Cart No. 1. With a self-satisfied air, he backed up to the nearest fire hydrant. Such devotion could hardly be ignored by city officials. Within a few days Old Gray was returned to the fire department by a less-than-satisfied furniture dealer, and restored to active duty as a reserve horse, always ready to make the run to a fire.[42]

The end of an era was coming. The old hook and ladder truck had been sold to a party in Reno in 1900; the Amoskeag steam fire engine that had belonged to Knickerbocker Engine Company No. 5 before the Great Fire of 1875 was sold to the fire department of Gilroy, California; and City Hose Cart No. 2 and the sleigh-hose-carriage were retired and sold to the Pierce A. Miller private collection in Modesto, California. Only City Hose Cart No. 1 and a skeleton crew of firemen, augmented by the volunteers in Divide Hose Company No. 2 and Liberty Engine Company No. 1, remained to keep watch over the Comstock. Even the ranks of the volunteers were beginning to thin. In Carson City, Warren Engine Company No. 1 became the first Nevada fire company to own a motorized fire engine, with its purchase of a 1913 Seagrave. On the Comstock, however, City Hose Cart No. 1 and the volunteers continued to respond to the fire alarm signal.[43]

By 1938, both the Divide Hose Company and Liberty Engine Company volunteers had formally disbanded and dissolved company funds and property. As a new generation of firemen sought to protect the Comstock, efforts were made to organize a new volunteer fire department for Virginia City and purchase a motorized fire truck. A 1929 Chevrolet cab and chassis was purchased by the fire department and delivered by the Virginia & Truckee Railroad. It was taken to the Overman Mine in Gold Hill, where Vic Maxwell, with the assistance of other volunteer firemen, built the Comstock's first motorized fire truck using materials Maxwell later said were "borrowed" from the mining company. The little truck proved its worth on many fires, but was itself almost destroyed by fire when the Consolidated Virginia works caught fire in 1939. Only driving the truck over some steep, rocky dumps saved it from destruction, according to former Chief Delbert Benner.[44]

Virginia City's fire service had at last entered the modern age. New volunteers with a new kind of fire engine were ready for any disaster, but the new equipment and new volunteers would soon find that fire has no regard for innovation and a new spirit. On November 13, 1942, fire broke out on the Divide between Gold Hill and Virginia City, posing the most lethal threat to the Comstock since the Great Fire of 1875. The fire broke out at about 9:30 p.m., on the road near the old Alpha dumps. Mrs. Amelia Sharon, a distant relative of the former mine owner and U.S. Senator, William Sharon, saw five youths on the road heading toward the Alpha dumps just west of her house. A short time later, as they hurried back down the road, she noticed a red tinge of flames licking about the sagebrush behind the homes near hers. The alarm was immediately sounded and the firemen rushed to the scene. The wind was blowing at a terrific rate and it fanned the flames into fury. The fire burned down behind the homes of Mollie Crocker and Will Greiner, but was stopped before it could do any damage to those residences.

The terrific wind swept the flames north and east across the Divide, though. Within a short time, the flames destroyed the homes of Matt Rheim, Tom Huddy, Lisa Kline, and Mrs. Sharon, sweeping with them a large boardinghouse, the home of Felix Romano, and the old firehouse of Divide Hose Company No. 2, west of C Street.

The United States had been at war for almost a year, and many of the Comstock's young men had already left for military service. It was soon evident that more than the resources of the tiny Virginia City Fire Department would be needed to stop the conflagration. Volunteers were sent from Dayton, Silver City, and Reno. Warren Engine Company No. 1 responded from Carson City, and Sparks Fire Chief Frank Hobson sent his auxiliary fire department to help. The Reno Police Department sent officers and a motorcycle squad to assist, along with deputies from the Washoe County Sheriff's Department. The State Grazing Division also had men in the field, and five trucks of U.S. Army troops from the encampment at Idlewild Park in Reno were soon on their way.

The flames had gained considerable headway, however. Crossing C Street and heading east, the fire soon consumed the homes of Frank Sharon, Julio Garavanta, and Tom Coyle, as well as the Lindsey and Clark residences. Flaming embers from the wind-whipped fire jumped across the ravine and started several spot fires near the Combination shaft. At one point it appeared to some that the old works were burning, but it proved to be some brush, old lumber, and a tree. A stand against the fire was attempted at the old C. T. Nutter store on the corner of Sheldon and C streets. For a time firemen thought they had saved the brick building with iron doors, which had been considered fireproof in the days when it was constructed. But a breach in the foundation allowed the flames to lick away at the old wooden floor, and the building was soon burning furiously from inside.

The fire moved rapidly, destroying a bar on the northeast corner of C and Sheldon streets, as well as the homes of Mrs. Tom Higgins, Charley Young, and John Tweedie and his sister. As the flames moved north toward Virginia City, residents began packing their belongings. Before long more than one hundred cars and trucks were ready to evacuate the city. The flames continued their march in two great fingers. One moved along South D Street and took with it the homes of Emie Pearl, William Nagle, Mamie Holland, Bill Pearson, Ben Wade, John Pittman, and the Johnson residence, as well as the old Lynds Grocery Store. The other finger moved along the base of the mountains towards the Con Chollar Mine pit. At one time it was thought the Fourth Ward School might burn, but the flames were kept from that building, as well as three other homes in the area.

The fire traveled swiftly up the mountain and across the top of the mine pit, towards the upper end of Virginia City, finally reaching homes above B Street. Another fire shot down Ridge Street towards the old

The 1942 fire was reminiscent of the Great Fire of 1875. Fanned by strong winds, the blaze swept over the Divide, destroying homes, businesses, and automobiles. Only a shift in the wind saved Virginia City from the same destruction that had taken place in 1875. (Walt Mulcahy)

Virginia & Truckee roundhouse. Fireman Elwood Johnson pulled up in front of the home of Enrico Giuffra and left hose and a nozzle and then quickly sped off. Within five minutes, the home of Bessie Gladding had virtually exploded in a mass of fire, while the Giuffra family applied wet burlap sacks to the roof of their house and sheds, keeping things as damp as possible with garden hoses and the fire hose that had been left them. The fire next consumed the homes of Andy Antunovich, Mandy Pappas, Eddie Bob, and the Corrales family.

On Howard Street, the smoke was so thick that reporter Ty Cobb, whose home was nearby on A Street, was driven from the phone while calling the story into the *Nevada State Journal* in Reno. The fire had gained headway about twenty feet into a house on Howard Street, and was moving across the mountain towards the Stoddard house when the wind suddenly shifted. This created a backfire situation and virtually extinguished the main body of the fire. The new generation of volunteers with their modern equipment had been unable to conquer the fire. It was stopped by the same winds that only minutes before had been driving it.

All during the night and for the next two days, volunteers, Army troops, and firemen quelled small spot fires and put out burning embers. When it was all over, the last "great fire" to visit the Comstock had laid waste to more than twenty-five homes and sheds, leaving more than a score of families homeless. In its fury, the fire had also burned a county truck, a large supply of sewer pipe, and a number of private cars and trucks. Only a handful of buildings on the Divide, once a populous and busy section of the Comstock, were spared. The homes that were consumed held not only the keepsakes of families, but also many historical artifacts, photographs, and records of the bonanza days of Virginia City and Gold Hill. It was a fire like that of 1875, which would leave a deep and lasting impression on the citizens and firemen of the Comstock.[45]

The Virginia Fire Department continued to grow and acquire newer, more modern equipment. It virtually abandoned the old Corporation House on the hill above B Street, after moving into the old American Saloon on C Street, which had been given up for back taxes. In 1948, at the urging of neighbors, the old Corporation House was razed by the county commissioners as a fire hazard, and City Hose Cart No. 1 was presented to the fledgling Nevada State Museum in Carson City. The passing of an era was complete, and on occasion an old-timer would lament the passing of the old paid fire department and the volunteers of the Divide and Gold Hill.[46]

In 1952, when heavy snows crushed the old Gold Hill hall of Liberty Engine Company No. 1, the four-wheel hose carriage and two-wheel hose cart inside were pulled from the wreckage and stored with the old carriage of Divide Hose Company No. 2. A group of modern volunteers then organized to build a monument to the pioneer fire fighters of the Comstock, using the old bell and tower that had once called the Liberty volunteers to fires.[47]

In 1976, a new generation of volunteer firemen embarked upon a serious project of restoration, calling themselves Liberty Engine Company No. 1, in honor of Gold Hill's longest-lived volunteer fire company. On July 4, 1979, the small band of volunteer firemen opened the Comstock Firemen's Museum in the old saloon that had housed the town's first motorized fire apparatus. In the tradition of the old volunteer companies, the occasion was a festive one, marked with a Fourth of July parade featuring virtually all of the fire department's modern equipment, as well as City Hose Cart No. 1, and the hose carriages of Divide Hose Company No. 2 and Liberty Engine Company No. 1. After the parade, these historic pieces of fire apparatus were housed in the traditional manner. The bell atop the firehouse museum sounded out a call to the people to join the firemen in "jollification," by enjoying the museum and refreshments as Virginia City citizens had done more than one hundred years before.

Within a year after the opening of the museum, this proud group of firemen began work on preserving the Exempt Firemen's Cemetery and organizing the first modern Comstock Firemen's Muster, with the cooperation of their Carson City friends in Warren Engine Company No. 1. The Storey County Commission also instigated legal action to have the 1929 Chevrolet pumper returned to Virginia City from neighboring Silver City, where it had been on loan since 1954, when no other motorized apparatus was available to that town for fire protection. Liberty Engine Company members continued to search for relics from the days of the old volunteer fire departments to display in the museum, and renewed many of the old traditions of the Comstock fire companies, in addition to starting a few of their own.

One of the finest traditions renewed was that of strong friendship with the volunteers of Warren Engine Company No. 1 in Carson City. This relationship was characterized by Warren Engine Company President Paul Webster as being unique in the modern fire service of Nevada for its extremely strong ties. The members of Liberty Engine Company No. 1 and the Warrens work side by side to organize traditional firemen's games and musters, meet socially, assist one another with special projects, and continue the traditions of friendship established by earlier generations of volunteer firemen in their respective companies. They compete in bucket brigade contests, hose cart races, and hand pumping contests with the same pride and determination of those who wore the red shirts and leather helmets in the early years of Nevada.

Tradition on the Comstock, it is said, is hard to kill.

Fireman Dave Tweedie: Making the Daily Run

DAVE TWEEDIE is perhaps the best remembered and most beloved fireman of the old Virginia Paid Fire Department. He was the driver of City Hose Cart No. 1 for many years.

Tweedie had been a member of the old volunteer fire department before creation of the paid department. He joined Monumental Engine Company No. 6, on the Divide, in 1874, and was one of the many firemen who fought the Great Fire of 1875.[1] After the volunteers disbanded following that fire, Monumental Engine Company No. 6 was allowed to remain in service, and Tweedie maintained his membership in that company until it finally disbanded. His skills as a teamster and stagecoach driver on the old Geiger Grade road between Virginia City and Reno were excellent qualifications for membership in the Virginia Paid Fire Department, and he decided to seek the position towards the turn of the century.

The late William "Bud" Spargo, who helped reel hose on the city hose carts as a boy, recalls that Tweedie stuttered when he got excited, which usually happened during a fire.[2] However, Dave Tweedie is best remembered for the way he kept his driving skills sharp, while also exercising the fire horses to keep them in shape for a quick run in the event of a fire alarm. Daily, at the sound of the noon chimes on the clock in the Corporation Fire House, Tweedie would tap the alarm bell, signalling the horses to fall into place underneath the hinged harness hanging in front of the hose carriage. He would hitch the team as fast as he could, as if there were a fire, and run them out of the firehouse up B Street, turning

Dave Tweedie on City Hose Cart No. 1, making his daily run on B Street. Tweedie is perhaps the best-remembered fireman from the old Virginia Paid Fire Department. (Dorothy Young Nichols)

down Carson Street to C Street on the north end of town, then galloping out to the Divide on the south end of town, where he would have lunch at his sister's house on Sheldon Street.[3]

Former Virginia City Fire Chief Delbert Benner recalls that, depending upon Tweedie's disposition at the moment, some of the local schoolboys would be allowed to ride on the back step of the hose carriage for at least part of the trip. If there were an alarm, he would yell for the youngsters to jump just before putting the horses to a dead run, rushing off to a chimney fire, stable blaze, or some other blaze threatening the town.[4]

Dave's return trip from his sister's house would be as fast and exciting as the trip to the Divide, and would generally reverse the original route, taking the horses and hose carriage back to the Corporation Fire House on B Street. However, Benner recalls that Tweedie also liked to return to the firehouse by driving along D Street, running City Hose Cart No. 1, with its gleaming red paint and shiny brass alarm gong, past the little houses in the redlight district and giving the girls on the line the thrill of seeing himself and the fine team strutting along. Perhaps there was a little gleam in his eye, too, as he waved to the girls on the porches and in the doorways of the little cribs.[5]

Tweedie left the fire department in the 1930s, when it became apparent there were changes in the wind and that the horse-drawn hose carriage might not be used for fire calls any longer. He moved to Napa, California, where he and his brother, John, owned the Pink Elephant Mine. Dave Tweedie is buried in the Napa Cemetery.[6]

Appendix 1

Chronology of the Comstock Volunteer Fire Companies

Company	*Date Organized*	*Disbanded*
Virginia Fire Department[a]		
1. Virginia Engine Co. No. 1	March, 1861	December, 1877
2. Nevada Hook and Ladder Co. No. 1	March, 1861	October, 1876
3. Young America Engine Co. No. 2	March 17/19, 1862	April 1, 1877
4. Eagle Engine Co. No. 3	August 2, 1863	November 9, 1875
5. Washoe Engine Co. No. 4	August, 1863	April 10, 1877
6. Knickerbocker Engine Co. No. 5	Summer, 1864	May 1, 1877
7. Confidence Engine Co. No. 6[b]	October 19, 1864	1879
8. Monumental Engine Co. No. 6[b]	May 11, 1866	1879
9. Hand-In-Hand Hose Co. No. 1	November 24, 1866	1866 or 1867
10. Lincoln Hose Co. No. 3	August, 1867	Unknown
11. Rooster Hose Co. No. 1	July, 1870	Unknown
12. Good Will Hose Co. No. 2	July, 1870	April 1, 1877
13. Invincible Hose Co. No. 4	July, 1870	March, 1871
14. Our Own Hose Co. No. 5[c]	1871	June 6, 1876
15. Neptune Hose Co. No. 5[c]	1871	June 6, 1876
16. Rescue Independent Hose Co. No. 1	1876	Unknown
Gold Hill Fire Department		
1. Silver Bar Hook & Ladder Co. No. 1[d]	November 18, 1863	August 19, 1864
2. Liberty Hose Co. No. 1[d]	August 19, 1864	1938
3. Yellow Jacket Hose Co. No. 2[e]	May 29, 1865	July 30, 1882
4. Lincoln Hose Co. No. 3	July, 1868	Unknown
5. Liberty Engine Co. No. 1[d]	July 28, 1868	1938
6. Yellow Jacket Engine Co. No. 2[d]	October 9, 1873	July 30, 1882
7. Divide Hose Co. No. 2	September 14, 1885	1937
8. Junior Hose Co. No. 1	Unknown	Unknown

a. After the Virginia Fire Department disbanded, the Virginia Exempt Firemen's Association was organized on November 26, 1876, and disbanded ca. 1915.

b. Confidence Engine Co. No. 6 was renamed Monumental Engine Co. No. 6 on May 11, 1866.

c. Our Own Hose Co. No. 5 was renamed Neptune Hose Co. No. 5 shortly after its organization in 1871.

d. Silver Bar Hook & Ladder Co. No. 1 disbanded on August 19, 1864, and reorganized as Liberty Hose Co. No. 1. With the arrival of a new hand engine, the company reorganized as Liberty Engine Co. No. 1 on July 28, 1868, retaining its hose company to run with the engine company.

e. Yellow Jacket Hose Co. No. 2 reorganized as Yellow Jacket Engine Co. No. 2 on October 9, 1873, after the arrival of a new steam fire engine. It retained its hose company to run with the new engine.

Appendix 2

Constitution and By-Laws of Eagle Engine Company No. 3 Virginia, Nevada. 1865.

REPORT

To the Officers and Members of Eagle Engine Co. No. 3:

Gentlemen: — Your Committee appointed to frame a Constitution and By-Laws for the government of the Company, report that they have performed the duty assigned them, and beg leave to submit the following for your consideration.

J. W. Hemenway,
Charles Rawson.

Thos. Peasley, Foreman.

CONSTITUTION.

ARTICLE I.

This Company shall be known and designated as Eagle Engine Company Number Three, and shall consist of not less than thirty-five, and not more than eighty members.

ARTICLE II.

The Company shall meet for the transaction of business on the first Monday of each month, at half past seven o'clock P.M. in the months of October, November, December, January, and March, and at eight o'clock P.M. in the months of April, May, June, July, August, and September. The annual meeting of the Company shall be held on the first Monday of August of each year, at eight o'clock P.M.

ARTICLE III.

The officers of the Company shall consist of a Foreman, First and Second Assistants, President, Secretary, and Treasurer, to be elected annually, by ballot, at the annual meeting of the Company in August, by a majority of the members present, and hold their office for the term of one year; in case of vacancy to be filled in like manner at the first monthly meeting thereafter. Three Delegates shall be elected in like manner at the stated meeting of the Company on the first Monday of June, to represent the Company in the Board of Fire Delegates of the Department, and shall hold their office for the term of one year.

ARTICLE IV.

Any citizen of the United States may be proposed for membership in this Company, when his name shall be referred to the Investigating Committee, who shall report thereon at the next stated meeting. If the report be favorable, the candidate may, upon paying an initiation fee of two and a half dollars and cost of certificate, be balloted for; and if five black balls be not cast against him, he shall become a member by signing the Constitution.

ARTICLE V.

It shall be the duty of the President to preside at all meetings of the Company; to call special meetings at the request of five members in writing, or whenever he may think proper. He shall also have the appointment of all committees not otherwise provided for.

ARTICLE VI.

It shall be the duty of the Foreman to see that the engine and apparatus are kept in good order and repair; in case of an alarm of fire, or when otherwise detailed for duty, to take command, and have a general supervision and control of the Company and apparatus during all drills, parades, and fires; to see that every member does his duty, and report delinquencies at stated meetings; and to enforce the constitution and bylaws of the Company.

ARTICLE VII.

It shall be the duty of the First Assistant Foreman to see that the engine is well manned and worked; that the members remain by the engine when on duty; to report to the Foreman all disobedience of orders, or neglect of duty on the part of the members; upon the return of the engine after a fire or alarm, to see that she is cleaned and in good order; to aid the Foreman in the discharge of his duties, and in his absence officiate in that capacity.

ARTICLE VIII.

It shall be the duty of the Second Assistant Foreman to take charge of the hose carriage and hose, and command the same when on duty. He shall appoint ten leading Hosemen, whose duty it shall be to assist him in the management of the same, and whose term of appointment shall be a period of three months. He shall order all members to remain by the engine not required by him, except those on special duty by order of the Foreman or First Assistant. He shall see that the hose carriage and hose are cleaned and in good order after a fire or alarm. He shall report to the Foreman all disobedience of orders or neglect of duty on the part of the members. In the absence of the Foreman and First Assistant, he shall have command of the Company.

ARTICLE IX.

It shall be the duty of the Secretary to keep a correct roll of the Company, and to call the same after every alarm of fire, or fire, and at the meetings of the Company. He shall keep a correct record of the proceedings of the Company in a book provided for that purpose, and at stated meetings read the minutes of the previous one. He shall keep a memorandum of all fires and alarms that may occur; collect all fines and dues from the Company, assessments or subscriptions, giving the person so paying a receipt, and pay the same to the Treasurer, taking his receipt therefor in a book to be kept for that purpose. He shall notify, in writing, every newly elected member of such election within five days after the time thereof, and furnish him with a copy of the code of laws; he shall, upon the receipt of his initiation fee and cost of certificate, make a return of such election to the Secretary of the Board of Delegates, take out his certificate of firemanship, furnish him with a key of the engine house, and place his name upon the roll of the Company. He shall report to the Secretary of the Board of Delegates all resignations, deaths, or expulsions that may occur in the Company, within five days after the same shall have occurred. On the second Monday of January, the second Monday of July, and ninety days preceding the first Monday of March, of each year, he shall furnish a complete return of the officers and members of the Company to the Secretary of the Board of Delegates, and compare the same with the record in his possession. He shall make quarterly reports to the Company of members in arrears for fines and dues, and shall call on such members and collect the same, which duty shall exempt him from all fines and dues.

ARTICLE X.

It shall be the duty of the Treasurer to receive all moneys collected by the Secretary; to give his receipt for the same; to pay all bills when duly audited by the proper committee, and which have been ordered paid by a vote of the Company; to keep a correct account of all receipts and disbursements in a book provided for that purpose; to submit his accounts to the Trustees when called upon by them; to make monthly reports to the Company of the receipts and disbursements, and no person shall be authorized to pay any bills except said Treasurer.

ARTICLE XI.

It shall be the duty of the Delegates to attend all meetings of the Board of Delegates of the Fire Department, and to report the proceedings of the same, in writing, at the following stated meeting of the Company.

ARTICLE XII.

It shall be the duty of each member to repair immediately to the engine upon every alarm of fire; to assist in drawing the same to and from the fire; to aid all in his power to get it to work; to remain where it is stationed; to assist in working the same, or to attend to any duty to which he may be detailed by any officer in command. It shall be the duty of the member first arriving at the engine house on an alarm of fire to take the trumpet, and act as Foreman during the absence of said officer or his assistants, and any member so acting shall be subject to all the penalties and privileges of the office he fills. It shall be the duty of each member to provide himself with a uniform of fire clothes and fire cap, in conformity with those worn by the Company, and a key of the engine house within thirty days after he shall have received his notice of election; to disperse from the engine house all persons under the age of twenty-one years, and to attend all meetings of the Company. No member shall be allowed to leave the engine when on duty, except by permission of the officer in command.

ARTICLE XIII.

Section 1. At the annual meeting of the Company in August, the President shall appoint a committee of three members, who shall be called the "Investigating Committee," whose duty it shall be to inquire rigidly into the character and competency of all candidates for membership after they shall have been proposed, and report in writing, using only the term "favorable," or "unfavorable," as the case may be; and their report shall not be questioned.

Sec. 2. At the annual meeting of the Company, in August, the President shall appoint a committee of three members, who shall be called "Trustees," whose duty it shall be to examine at least twice a year into the books and accounts of the Secretary and Treasurer, and report to the Company in writing; and attend to such other business as may from time to time be laid before him.

ARTICLE XIV.

HONORARY MEMBERS.

There shall be, and there is hereby established, an honorary roll of the Company, subject to the following rules and regulations: —

Section 1. Any person of good standing in the community may be proposed for honorary membership; and upon being balloted for, if he receives the unanimous vote of the Company present, at a stated meeting, he may become an honorary member, upon his paying to the Secretary the sum of one hundred dollars, which sum shall exempt him from all further assessments.

Sec. 2. Any officer or member of this Company, or member of the Fire Department, who shall have served faithfully the term of years to entitle him to exemption — or who shall have at a fire or otherwise, when on duty as a fireman, distinguished himself in a manner rendering the act meritorious — may be elected an honorary member by the unanimous vote of the Company present at a stated meeting.

Sec. 3. Any officer or member of the Company who has left or intends leaving the city, with a view to a permanent residence in some other county or State, and who, while a member, did his duty faithfully, may also be entitled to election as an honorary member, by the unanimous vote of the Company present at a regular meeting.

Sec. 4. There shall be engraved for the use of the Company, a certificate entitled "A Certificate of Honorary Membership of Eagle Engine Company, No. 3;" which certificate, on presentation, shall be signed by the Foreman, Secretary and Treasurer, and bear the seal of the Company.

Sec. 5. It shall be the duty of the Secretary to furnish each member elected to honorary membership, with a certificate, within five days of his qualification as such.

ARTICLE XV.

CONTRIBUTING MEMBERS.

There shall be and is hereby established, a contributing roll of the Company, subject to the following rules and regulations.

SECTION 1. Any person of good standing in the community may be proposed for contributing membership; he shall be balloted for in the same manner as active members at a stated meeting; when, if elected, he may become a contributing member upon his signing the Constitution and paying to the Secretary the sum of ten dollars, and which sum he shall pay semi-annually thereafter; and any member failing so to pay within thirty days after the same is due, shall be liable to expulsion.

SEC. 2. The members of the contributing roll shall have full privilege to attend any meeting of the Company and take part in all discussions, but shall not be entitled to vote. They shall be held exempt from fire duty, but shall be allowed to wear the uniform of the Company and assist in rolling the engine to and working at fires. When in uniform, they shall be subject to the same orders and penalties for violation of the same, as in the case of active members. They shall have power to roll the engine, in case of fire, and shall have command until the arrival of an active member. No person being a member of another fire company, shall be elected without resigning from said other Company.

SEC. 3. There shall be engraved, for the use of the company, a certificate entitled "A Certificate of Contributing Membership of Eagle Engine Company, No. 3;" which certificate, on presentation, shall be signed by the Foreman, Secretary and Treasurer, and shall bear the seal of the Company.

SEC. 4. It shall be the duty of the Secretary to furnish each member elected to contributing membership, with a certificate, within five days after his qualification as such.

ARTICLE XVI.

All moneys received from contributing members, shall be denominated a "Charitable Fund," from which shall be paid no moneys, except ordered by a special vote of the Company, on which the ayes and noes shall be taken.

ARTICLE XVII.

IMPEACHMENT.

SECTION 1. The Company, when called together by a request of any ten members, shall have power to impeach the Foreman, or any of the Assistants, or any one holding office in this Company, for neglect or inability to discharge the duties of his office, or for an unwarrantable exercise of power, or for drunkenness while in the discharge of the duties of his office. And when an impeachment shall be made, the party impeached shall, after receiving one week's notice of the same, together with a written specification of the charges made, be tried before the Company; and shall be liable to such action as the Company may deem proper.

SEC. 2. All charges and specifications of charges shall be preferred in writing to the Company, in the name of Eagle Engine Company, No. 3; a copy of which shall be furnished the accused at least three days before trial. Charges and specifications of charges may be preferred at any meeting, stated or special, when the Company shall forthwith set a time for trial, which shall not be within seven days from the date the charges may be preferred, unless by consent of both parties. All witnesses shall be examined in the following order: 1st, prosecution; 2nd, defense. Rebutting evidence may be offered on either side; and when the testimony shall have been closed, the Company shall, without debate, proceed to find upon the guilt or innocence of the accused. If guilty, the Company shall then assess the punishment, which shall be, removal from office, fine, suspension, or expulsion, at the option of the Company. In all cases the President may appoint one member of the Company to conduct the prosecution, and the accused one to conduct his defense.

ARTICLE XVIII.

This Constitution may be amended by a two third vote of the Company present, at any regular meeting, provided one month's notice thereof be given in writing.

BY-LAWS.

ARTICLE I.

ROLL CALL.

The Secretary shall call the roll at every meeting of the Company, and after every alarm of fire.

ARTICLE II.

In case of fire the first two members arriving at the engine house shall be entitled to the pipes.

ARTICLE III.

Any officer of the Company, or any member of the Investigating or Standing Committees, shall, for any neglect of duty, incur a fine of not less than two dollars, nor more than five dollars, at the discretion of a majority of the Company.

ARTICLE IV.

FINES OF MEMBERS.

SECTION 1. For each non-attendance at stated meetings, one dollar.

SEC. 2. For leaving a meeting without permission of the Chairman, fifty cents.

SEC. 3. For refusing to come to order when called upon, one dollar.

SEC. 4. For absence at roll call upon the return of the engine after a fire, when she has been worked, one dollar.

SEC. 5. For absence at roll call upon the return of the engine after a fire or alarm, when not worked, fifty cents.

SEC. 6. For leaving the engine when on duty without permission of the officer in command, one dollar.

SEC. 7. For leaving the drag rope or forbearing assistance, one dollar.

SEC. 8. For disobedience of orders, five dollars.

SEC. 9. For any violation of the rules and regulations for the government of the meeting room, one dollar, and to be collected forthwith. with.

SEC. 10. For introducing any religious or political subjects, or using profane language at the meetings of the Company; for making known to any person not a member, any remarks made at a meeting, or divulging the business of the Company, a member shall incur a fine of five dollars, and for a repetition of either offense be liable to expulsion.

SEC. 11. Each member shall contribute one dollar monthly to defray the contingent expenses of the Company.

SEC. 12. All fines and dues must be settled monthly.

SEC. 13. For not appearing at drill, each member so absent shall be fined two dollars.

ARTICLE V.

PLEAS.

Sickness or death in the family of a member; personal sickness; absence from the city, or being at work, are the only excuses which can be received for any neglect of duty, except by the assent of a majority of the members present at a stated meeting. Any one giving an incorrect excuse, shall be fined or expelled at the discretion of the Company.

ARTICLE VI.

RESIGNATIONS.

No resignation of a member shall be accepted until all fines, dues and penalties are discharged, except by a two third vote of the members present.

ARTICLE VII.

EXPULSION.

SECTION 1. Whenever a member is repeatedly deficient in the discharge of his duty, or does not exert himself to arrive at the engine, he shall be notified by the Secretary to appear at the next monthly meeting of the Company; if he fails to appear, or his excuse be deemed insufficient by the Company, he shall be expelled upon a motion from any member.

SEC. 2. If a member neglects three stated meetings of the Company, or three roll calls, in succession, without sending a written excuse; if he be known to give an incorrect excuse for delinquency; if he use insulting language towards an officer of the Company; if he be considered an improper associate, he shall be expelled.

SEC. 3. Any member who neglects to pay his fines, dues, and assessments for three months, may be expelled.

SEC. 4. All expulsions shall be by ballot, and shall require a two third vote of the Company present.

SEC. 5. The Secretary shall notify the Foreman of each Company in the department and the Secretary of the Board of Delegates, of each expulsion made from the Company, within five days of the time thereof, stating the reason why such expulsion was made.

ARTICLE VIII.

ORDER OF BUSINESS.

SECTION 1. The order of proceedings shall be as follows, ten members constituting a quorum.

1. Roll Call.
2. Reading Minutes of Previous Meeting, and Question on Approval.
3. Collection of Fines, Dues and Assessments.
4. Report of Treasurer and Delegates.
5. Report of Special Committees.
6. Report of Investigating Committee.
7. Resignations and Expulsions.
8. Election of Officers and Members.
9. Unfinished Business.
10. Miscellaneous Business.
11. Roll Call.
12. Adjournment.

SEC. 2. The Chairman shall preserve order and decorum; appoint committees of three or under, with the consent of the Company; and shall have none but a casting vote, except when more than a majority is required, or, when voting by ballot; he shall take no part in debate while in the Chair.

SEC. 3. A member wishing to speak on any question, shall rise and address the Chair, and if two or more members rise at the same time, the one furthest from the Chair shall be deemed to have the floor.

SEC. 4. No member shall speak more than twice on the same question, nor for more than five minutes at a time, unless by permission of the Chair.

SEC. 5. No debate shall be heard on any motion until the same shall have been seconded, and when a question is before the Company, no motion shall be in order, except—

1. To lay on the table;
2. To amend;
3. To postpone;
4. To reconsider;
5. To refer;

6. The previous question;
7. To adjourn;

and they shall take precedence in the order in which they are arranged—the last two to be decided without debate.

SEC. 6. When any question is put, every member present shall vote either for or against the same, unless excused by the Chair.

SEC. 7. After any question has been decided, any two members who voted in the majority, may move for a reconsideration thereof, but no discussion of the main question shall be allowed, unless the same has been reconsidered.

SEC. 8. The Chair shall state every question coming before the Company, and, immediately before taking the vote, shall ask: Is the Company ready for the question? and should no member rise to speak, he shall take the question, and after it has been taken no member shall speak upon it, unless by consent of the Company. He shall pronounce the vote and decisions of the Company upon all subjects.

SEC. 9. A motion to adjourn shall always be in order, and shall be decided without debate; if no time be specified, it shall be until the next regular meeting.

SEC. 10. Any member may appeal from the decision of the Chair on a point of order. The question shall be—Do you sustain the Chair? and it shall be decided by a majority, without debate.

SEC. 11. A motion to take the previous question may be made by any two members, and shall be put in this form: Shall the main question be now taken? and if adopted, all further debate, and all amendments which have not been adopted, shall be excluded, and the question shall be taken on the original motion.

SEC. 12. The yeas and nays shall not be taken on any question unless called for by at least four members.

SEC. 13. Any member may call for a division of the question, when the sense will admit of it, and a majority may decide whether it shall be divided or not.

SEC. 14. When it is voted to lay any subject on the table, it cannot be taken up at the same meeting, unless by a two third vote of the members present.

SEC. 15. Any section of Article Eight may be suspended by a two-third vote of the members present, such suspension to terminate with the meeting.

ARTICLE IX.

GOVERNMENT OF MEETING ROOM.

SECTION 1. The meeting room shall not be used for any purpose except for the transaction of the business of the Company, except by the consent of the Foreman, and a majority of the Company present at a meeting.

SEC. 2. Any member who shall be guilty of intentionally defacing the walls of the engine house, by writing upon them or otherwise; or removing any property of the Company from the house; or of intentionally injuring the furniture of the rooms, shall be liable to expulsion.

SEC. 3. No smoking shall be allowed in the meeting room, under a penalty of one dollar for each offense.

SEC. 4. At every meeting of the Company the Chairman shall appoint a Sergeant-at-Arms, who shall be stationed near the entrance of the room, and whose duty it shall be to report to the Secretary all members arriving after roll call; to report all members leaving the meeting without permission of the Chair; and to report all violations of Sections second and third of this Article to the Foreman; such appointment to terminate with the meeting.

ARTICLE X.

No officer shall be allowed to contract any debt for the Company to exceed in amount the sum of fifty dollars, except permission be granted by vote at a meeting of the Company, and any officer so doing shall be individually liable therefor.

ARTICLE XI.

No officer or member of the Company shall loan any part or portion of the paraphernalia of the Company without the permission of five members and the concurrence of one officer signed to a written permission.

ARTICLE XII.

These By-Laws may be amended by a two third vote of the Company present at any regular meeting; provided that one month's notice thereof be given in writing.

Appendix 3

Virginia City's Great Fire October 26, 1875

Time fire started: Approximately 5:30 a.m.

Origin of fire: Upsetting of a lamp in the basement of a small lodging house at No. 19 South A Street, owned by "Crazy Kate" Shea.

Area destroyed: Approximately 33 square blocks; south to just beyond Taylor Street, north beyond Carson Street, east to below G Street, and west to Stewart Street.

Losses: More than 1,000 homes. More than 300 commercial establishments (virtually the entire business district). Prominent structures destroyed included: the Virginia & Truckee Railroad depot and facilities, Piper's Opera House, city offices and jail, county courthouse, St. Mary's in the Mountains Catholic Church, St. Paul's Episcopal Church, the Methodist Church, *Virginia Evening Chronicle* offices, *Territorial Enterprise* offices, Bank of California, most of the city's fraternal lodges, Virginia Savings Bank, Consolidated Virginia Mill, California Battery Mill, Ophir Mine and Mill buildings, International Hotel, and most of the saloons of the city.

Deaths: 3, all from falling walls.

Financial loss: More than $20,000,000 ($12,000,000 in commercial establishments).

Fire Department losses:

Eagle Engine Company No. 3, chemical extinguisher destroyed	$3,000
Neptune Hose Company No. 5, cart destroyed	300
Monumental Engine Company No. 6, cart destroyed	250
Knickerbocker Engine Company No. 5, steamer damaged, suction lost	300
Eagle Engine Company No. 3, engine house destroyed	3,000

Knickerbocker Engine Company No. 5, engine house destroyed	3,000
Virginia Engine Company No. 1, engine house destroyed	2,000
Nevada Hook & Ladder Company No. 1, engine house destroyed	1,500
2,500 feet of carbolized hose destroyed	1,500
1,000 feet of leather hose destroyed	1,000
9 cisterns damaged or destroyed	180
Yellow Jacket Engine Company No. 2, steamer check valve ruined	20
Total	$16,050

In addition, Nevada Hook and Ladder Company No. 1 lost all ladders, hooks, and other tools and equipment; all companies lost considerable numbers of tools and equipment; No. 3, No. 5, No. 1, and the hook and ladder company lost all paraphernalia in their engine houses, as well as their records. Also lost was the fire bell of Eagle Engine Company No. 3.

Active roll of the Virginia Fire Department prior to the fire: More than 500.

Active roll of the Virginia Fire Department after the fire: Fewer than 250.

Appendix 4

Glossary

Amoskeag—The Amoskeag Company was located in Manchester, New Hampshire, and began building steam fire engines in 1859. Some 854 Amoskeag engines were sold to fire departments by Amoskeag and its successor companies.

Babcock—The Babcock Manufacturing Company of Chicago produced chemical fire-fighting apparatus which employed a mixture of bicarbonate of soda, sulfuric acid, and water, creating a pressurized gas (carbon dioxide) which was used to extinguish fires. These wagons were first pulled by hand, but by 1872, horse-drawn chemical engines were being produced.

Brakes—The rods or bars by which the pump on a fire engine was manually operated.

Butt—The end of the hose to which the pipe or nozzle was attached, usually made of brass.

Butt Enders—Nickname given to members of Gold Hill's Liberty Engine Company No. 1, probably because the engine house of the company was located at the upper or butt-end of Gold Hill.

Button & Blake—Lysander Button gave his name to the company that would build some 500 fire engines, beginning in 1834, in a plant located in Waterford, New York. At various times the company was known as Button & Son, Button & Blake, and the Button Manufacturing Company. Button also produced hose carriages, hose carts, and other equipment.

Carbolized hose—An early, high-pressure hose used with steam fire engines.

Cistern—A large water tank dug into the ground, lined with wood or stone, and caulked, from which fire engines drafted water.

Clapp & Jones—This company eventually produced 400 steam fire engines in its plant in Hudson, New York, and was one of three companies to dominate the steam fire engine market after 1891.

Company—A group of firemen organized to combat fire. Examples were hook and ladder companies, engine companies, and hose companies. See Department.

Crib—A small one- or two-room shack in which a prostitute lived and worked. A row of cribs was known as "the line."

Department—The parent organization to which companies belonged. It oversaw the operations and conduct of the individual companies and their members.

Engine—A fire engine or hand pumper. Engine was often pronounced with emphasis of the "i" by early firemen and observers, and was popularly spelled "enjine."

Exempts—Members of an exempt firemen's organization, usually comprised of former members of a volunteer fire department who had served a minimum number of years, or who were physically unfit to respond to fires. Exempt members were exempt from responding to fires and from attending meetings and drills.

Fire jakies—Nickname given to volunteer firemen. Origin is obscure.

Fire laddies—Nickname given to volunteer firemen.

First water—A fireman's battle cry, shouted when his company was the first to put water on a fire, signifying the quickness and efficiency of that company.

Hand engine—A fire engine pumped manually by firemen working the "brakes."

Hooks—Nickname given to a hook and ladder company.

Hubbs' babies—Large two-wheeled hose carts that could be pulled by hand or behind an engine. Named after New York fireman David Hubbs, who designed them.

Hunneman—Hunneman & Company, located in Boston, built 716 hand-operated fire engines. These engines were popular because of their light weight and mobility.

Incendiary—An arsonist.

Jackets—Nickname given to members of Gold Hill's Yellow Jacket Engine Company No. 2.

Jakie—See Fire jakie.

Jeffers—A Rhode Island manufacturer of hand engines and steam fire engines, run by William Jeffers.

Jumper—A two-wheel hose cart capable of carrying several hundred feet of hose. Also known as a "Hubbs' baby," or a "tender." The jumpers varied in size and capacity and some were modified to be pulled behind a steam fire engine or hand pumper. Quite often jumpers were used in hose cart races.

Machine—A hand pumper, popularly spelled "masheen."

Markers—Flags, usually silk, bearing the name and/or number of a fire company. Such flags were usually used in parades.

Muster—A gathering of firemen and fire apparatus in traditional contests such as hose cart races and hand pumping contests.

Pipe—A nozzle.

Play—Guide a stream of water.

Pumper—A hand engine.

Salamander—Nickname for a fireman, popularly supposed to resist fire or heat, as a salamander.

Silsby—One of the earliest manufacturers of steam fire engines was the Silsby Manufacturing Company of Seneca Falls, New York. Their pumps employed a rotary design as opposed to the piston or reciprocating pumps used by other manufacturers. Silsby also produced hand engines.

Spanner—A wrench or tool used for coupling hose.

Squirrel tail—Name given to hand engines that featured a large tube attached over the top of the engine, in which the suction hose was stored.

Squirt—The act of squirting water through a hose from a hand pumper. A contest or a trial was often called "a squirt." The name also applied to the distance attained by the water pumped from a hand pumper.

Steamer—A steam fire engine.

Steward—A member of a fire company elected to act as caretaker of the engine house and the property of the engine company. Stewards received a salary and often lived in the engine house. Duties might also include keeping a fire burning in the stove to prevent the engine from freezing in the winter.

Truck—The hand-drawn or horse-drawn wagon which carried the hooks, ladders, and other equipment of a hook and ladder company.

Tub—Nickname given to Hunneman hand engines because of the large tub surrounding the pump, which held water. This differed from the small tanks on other engines, which only held enough water to "prime" the pump.

Turn tongue in—When an engine company was suspended, or went on strike, it was said to be "turned tongue in," and therefore would not respond to fires. In some cases, apparatus was housed with the tongue away from the doors, so as to be more difficult to pull out.

Washing—When two teams "faced off" in a squirting contest, the engine which doused the other with the most water was said to be washing it. Also, in another kind of contest, one machine would draw water from a cistern as rapidly as possible, pumping it into the tank of a second one; men operating the second machine would try to pump fast enough to keep their tank from overflowing; if the water overflowed and ran down the sides of the second machine, it was being washed.

Washoe Seeress—Nickname given to Eilley Orrum Bowers, who claimed to be clairvoyant and told fortunes utilizing a crystal ball.

Notes

Chapter I
From Snowballs to Hand Pumpers

1. Mark Twain, *Roughing It* (New York: American Library of World Literature, Inc., 1962), pp. 227–228.
2. *Ibid.*, p. 228.
3. Irving Stone, *Men To Match My Mountains* (New York: Doubleday & Co., 1956), p. 209; Eliot Lord, *Comstock Mining and Miners* (San Francisco: Howell-North Books, 1959), p. 66.
4. C. B. Glasscock, *The Big Bonanza* (Indianapolis: Bobbs-Merrill Company, 1931), p. 71.
5. Twain, p. 229.
6. San Francisco *Alta California*, November 8, 1861.
7. J. Wells Kelly, *1863 Directory of Nevada Territory*, Private Collection, Virginia City, Nevada; *Grand Register of the Virginia Fire Department*, Comstock Firemen's Museum, Liberty Engine Company No. 1, Virginia City, Nevada.
8. *Constitution and By-Laws, Eagle Engine Company No. 3* (San Francisco: 1865), Nevada State Museum, Carson City.
9. *Grand Register of the Virginia Fire Department*.
10. *Gold Hill News*, January 24, 1865.
11. *Minutes, Gold Hill Fire Department*, May 31, 1878. Comstock Firemen's Museum, Liberty Engine Company No. 1, Virginia City, Nevada.
12. *Constitution and By-Laws, Virginia Fire Department* (San Francisco: H. S. Crocker & Co., March 13, 1872), Nevada Historical Society, Reno.
13. *Constitution and By-Laws, Eagle Engine Company No. 3*.
14. Alfred Doten, *The Journals of Alfred Doten, 1849–1903*. Ed. by Walter Van Tilburg Clark. (Reno: University of Nevada Press, 1973), p. 917.
15. *Virginia Daily Union*, March 5, 1865; *Gold Hill News*, March 6, 1865.
16. *Territorial Enterprise*, April 29, 1897.
17. Myron Angel, *History of Nevada, 1881* (Berkeley, California: Howell-North Books, 1958; Oakland: Thompson & West, 1881), p. 600; Doten, pp. 881, 981–982, 1037; *Grand Register of the Virginia Fire Department*.
18. *Ibid.*
19. *Minutes, Gold Hill Fire Department*.
20. *Ibid.*; Doten, pp. 880, 881.
21. *Laws of the Virginia Fire Department* (Virginia City: Enterprise Steam Press, 1864), Comstock Firemen's Museum, Liberty Engine Company No. 1, Virginia City, Nevada.

22. *Gold Hill News*, February 21, 1865.
23. Doten, p. 836.
24. *Gold Hill News*, June 1, 1865.
25. *Ibid.*, June 3, 1865.

Chapter II
Dangerous Work

1. *Gold Hill News*, September 10, 1869; see also *Territorial Enterprise*, September 10, 1869.
2. *Gold Hill News*, September 10, 1869.
3. *Ibid.*, August 12, 1868.
4. *Ibid.*, January 12, 1869.
5. Doten, p. 825.
6. *Gold Hill News*, April 18, 1864.
7. *Virginia Daily Union*, August 8, 1865; *Gold Hill News*, August 7, 1865.
8. *Gold Hill News*, November 13, 1868.
9. *Ibid.*, September 10, 1869.
10. *Ibid.*, December 26, 1864.
11. *Ibid.*, January 26, 1864.
12. Doten, p. 832; *Gold Hill News*, April 28, 1865.
13. *Gold Hill News*, August 22, 1867.
14. Doten, p. 956; *Gold Hill News*, October 26, 1867.
15. *Gold Hill News*, February 8, 1864.
16. *Ibid.*, October 29, 1872.
17. *Ibid.*

Chapter III
Gold Hill's Great Fire of 1870

1. *Gold Hill News*, July 5–7, 1870; Doten, pp. 1097–1098.
2. *Gold Hill News*, June 13–15, 1871; Doten, p. 1132.

Chapter IV
The Social Life of a Fireman

1. Doten, pp. 910, 911.
2. *Gold Hill News*, February 24, 1872.
3. *Ibid.*, March 14, 1873.
4. *Territorial Enterprise*, July 6, 1871; *Gold Hill News*, July 5, 1871.
5. Doten, pp. 1012–1013.
6. *Gold Hill News*, September 6, 1870.
7. *Ibid.*, December 28, 1869.
8. *Ibid.*, June 17, 1869.
9. *Territorial Enterprise*, July 6, 1869; *Gold Hill News*, July 6, 1869.
10. *Nevada State Journal*, February 22, 1934.

11. *Gold Hill News*, June 20, 1871.
12. *Territorial Enterprise*, May 27, 1870; *Gold Hill News*, May 26, 1870.
13. *Gold Hill News*, August 6, 1868.
14. *Roll Book, Gold Hill Fire Department*, Comstock Firemen's Museum, Liberty Engine Company No. 1, Virginia City, Nevada.
15. *Gold Hill News*, July 5, 1868.
16. *Ibid.*, December 6, 1867.
17. *Ibid.*, December 26, 1867.
18. *Ibid.*, March 31, 1864.
19. Douglas McDonald, *The Legend of Julia Bulette* (Las Vegas, Nevada: Nevada Publications, 1981).
20. Doten, pp. 912, 932–934.

Chapter V
Home Sweet Home

1. *Gold Hill News*, November 7, 1863.
2. *Ibid.*, September 29, 1864.
3. *Ibid.*, October 4, 1869.
4. *Ibid.*, June 13, 1871.
5. *Nevada State Journal*, August 18, 1927.
6. *Virginia Daily Union*, July 4, 1865; *Gold Hill News*, July 3, 1865.
7. *Gold Hill News*, February 10, 1873.
8. *Ibid.*, November 11, 1869.
9. *Ibid.*, August 31, 1875.
10. *Virginia Evening Chronicle*, November 2, 1879.
11. *Gold Hill News*, August 12, 1867.
12. *Territorial Enterprise*, February 3, 1871; *Gold Hill News*, February 3, 1871.
13. *Gold Hill News*, April 7, 1870.
14. *Ibid.*, June 26, 1867.
15. *Territorial Enterprise*, July 13, 1877, June 20, 1878.
16. *Gold Hill News*, June 17, 1869.
17. *Ibid.*, July 8, 1869.
18. *Ibid.*, January 17, 1876.
19. *Virginia Evening Chronicle*, November 2, 1879.
20. Doten, p. 1378.
21. *Gold Hill News*, March 7, 1876.
22. *Minutes, Virginia Board of Aldermen*, December 7, 1877, Storey County Clerk's Office, Virginia City, Nevada.

Chapter VI
Scandal

1. *Virginia Daily Union*, January 24, 30, 1865; *Gold Hill News*, January 23, 30, 1865.
2. *Gold Hill News*, January 23, 30, 1865.
3. *Minutes, Gold Hill Fire Department*, August 13, 16, 17, September 10, 1886.

4. *Ibid.*, April 30, 1889.

5. *Minutes, Liberty Engine Company No. 1,* August 5, September 22, 1910, Comstock Firemen's Museum, Liberty Engine Company No. 1, Virginia City, Nevada.

6. *Ibid.*, February 8, March 8, 1927, August 12, 1930.

7. *Carson Daily Appeal,* December 30, 1887.

Chapter VII
Rivalries and Friendships

1. Doten, p. 840; *Gold Hill News,* June 29, 1865.

2. *Minutes, Liberty Engine Company No. 1,* June 1, 1870; *Gold Hill News,* June 6, 10, 23, 1870; Doten, p. 1095.

3. *Gold Hill News,* December 18, 19, 20, 1865.

4. *Ibid.*, August 22, September 27, 1865.

5. *Ibid.*, October 2, 1865.

6. *Ibid.*, August 16, 1865.

7. Doten, pp. 879, 881–882.

8. *Gold Hill News,* March 3, 1869.

9. *Ibid.*, April 5, 1872.

10. *Ibid.*, May 21, 1874.

11. *Minutes, Liberty Engine Company No. 1,* June 25, 1896.

12. *Ibid.*, September 21, 1897.

13. *Ibid.*, June 13, 1899.

14. *Gold Hill News,* August 21, 1876.

15. *Grand Register of the Virginia Fire Department.*

16. Angel, p. 599.

17. *Ibid.*, p. 598.

18. *Ibid.*, p. 599; *Gold Hill News,* November 4, 5, 1863.

19. *Gold Hill News,* November 6, 1863.

20. *Ibid.*

21. *Virginia Daily Bulletin,* February 11, 1864; *Gold Hill News,* February 11, 1864; *Territorial Enterprise,* March 22, 1864; *Virginia Daily Union,* March 22, 1864.

22. Ibid., *Grand Register of the Virginia Fire Department.*

Chapter VIII
Man the Brakes!

1. *Gold Hill News,* November 29, 1872.

2. *Ibid.*, March 24, 1873.

3. *Angel,* p. 588.

4. *Gold Hill News,* January 30, 1865.

5. Doten, p. 832.

6. *Ibid.*, p. 905.

7. *Minutes, Virginia Board of Aldermen,* January 15, 29, 1878.

8. *Territorial Enterprise,* December 13, 1877, January 15, 1878.

9. *Minutes, Virginia Exempt Firemen's Association*, June 12, 1887, Nevada Historical Society, Reno, Nevada.

10. *Ibid.*, March 8, 1885; *Virginia Evening Chronicle*, December 22, 1896; Scrapbook, Gardnerville Volunteer Fire Department, Bud Brown, Gardnerville, Nevada.

11. Charles Collins, *Mercantile Guide & Directory for Virginia City, Gold Hill, Silver City and American City, 1864*–1865, Special Collections Department, University of Nevada-Reno Library, Reno, Nevada; Angel, p. 588.

12. *Gold Hill News*, December 21, 1863; Angel, p. 588.

13. *Gold Hill News*, June 19, 1865.

14. Doten, p. 956.

15. *Territorial Enterprise*, July 25, 1872.

16. Angel, p. 588, *Virginia Evening Bulletin*, October 23, 1863.

17. *Gold Hill News*, June 19, 1865.

18. *Ibid.*, June 10, 14, 1870.

19. *Ibid.*, February 11, 1865.

20. *Ibid.*, October 26, 27, 1875.

21. *Ibid.*, November 9, 1875.

22. Angel, p. 588; *Virginia Evening Bulletin*, October 23, 1863; *Gold Hill News*, November 3, 1863.

23. *Gold Hill News*, November 7, 1863.

24. *Ibid.*, November 30, 1863.

25. *Ibid.*, July 31, 1865.

26. *Territorial Enterprise*, May 6, 1869; *Gold Hill News*, June 29, 1872.

27. *Gold Hill News*, August 29, 1872; Doten, p. 1173.

28. *Gold Hill News*, March 9, 1876; *History of the Reno Fire Department*, compiled by Allen M. Robinette (no place, no publisher, 1908).

29. *Gold Hill News*, June 12, 1865, September 14, 1869.

30. *Ibid.*, July 29, August 7, 15, 1870.

31. *Territorial Enterprise*, August 27, 1872; *Gold Hill News*, July 13, August 28, 1872.

32. *Gold Hill News*, March 17, 1865.

33. *Ibid.*, April 26, May 1, 1865.

34. *Ibid.*, May 12, 1873.

35. *Virginia Evening Chronicle*, April 10, May 5, 1885.

36. *Territorial Enterprise*, June 28, 1868.

37. *Gold Hill News*, September 26, 1868.

38. Doten, p. 1045; *Gold Hill News*, April 28, May 5, 6, 7, 10, 1869.

39. *Ibid.*, September 12, 1872, November 13, 1873.

40. *Ibid.*, March 26, 31, 1873.

41. *Minutes, Liberty Engine Company No. 1*, March 13, 1894.

42. *Ibid.*, September 22, 1910.

Chapter IX
The Hooks

1. Collins, *Mercantile Guide & Directory for Virginia City, Gold Hill, Silver City & American City*, 1864–1865.

2. *Gold Hill News,* December 7, 1863; *Grand Register of the Virginia Fire Department.*

3. *Gold Hill News,* August 1, 1864.

4. Henry Nash Smith, *Mark Twain of the Enterprise* (Berkeley: University of California Press, 1957).

5. Angel, p. 266.

6. *Gold Hill News,* January 13, 1869.

7. *Territorial Enterprise,* August 30, 1869; *Gold Hill News,* August 30, 1869.

8. *Gold Hill News,* May 9, 1870.

9. *Ibid.*, November 10, 1875, March 9, 1876.

10. *Ibid.*, March 9, 1876; *Territorial Enterprise,* March 15, 1876; *Minutes, Virginia Board of Aldermen,* October 10, 31, November 1, 1876.

11. *Virginia Evening Chronicle,* January 26, 1900.

Chapter X
Jump Her Lively, Boys!

1. *Territorial Enterprise,* July 12, 14, 25, 1870; *Gold Hill News,* July 12, 15, 26, 1870, July 1, 1871; Doten, p. 944.

2. Paul C. Ditzel, *Fire Engines—Firefighters* (New York: Crown Publishers, 1976), p. 61.

3. Doten, p. 832.

4. *Gold Hill News,* September 12, 1876.

5. *Ibid.*

6. *Ibid.*, June 10, 1871.

7. *Ibid.*, August 24, September 7, 1870, July 12, 1871, July 6, 1872.

8. *Territorial Enterprise,* June 7, 1876.

9. *Gold Hill News,* August 15, 22, 1864.

10. *Ibid.*, October 24, 1864.

11. *Ibid.*, June 24, 1865.

12. *Ibid.*, June 28, July 1, 1865.

13. *Ibid.*, September 7, October 26, 1870.

14. *Territorial Enterprise,* November 14, 1870; *Gold Hill News,* November 14, 1870.

15. *Minutes, Liberty Engine Company No. 1,* October 30, 1880; Doten, pp. 1372–1373.

16. *Nevada State Journal,* February 22, 1934.

17. *Minutes, Yellow Jacket Engine Company No. 2,* April 9, 1879, Nevada Historical Society, Reno, Nevada.

18. *Territorial Enterprise,* October 2, 1885; *Minutes, Board of Delegates of the Gold Hill Fire Department,* September 10, 21, October 11, 1886.

19. Richard C. Datin, *Elegance on C Street* (Reno, Nevada: n.p., 1977), p. 43.

20. Interviews with Delbert Benner, 1979–1980.

21. Interviews with Delbert Benner, Jesse Young, William "Bud" Spargo, Alice Byrne, Vic Maxwell, 1975–1980.

22. *Territorial Enterprise,* March 22, April 26, May 19, 1877; *Minutes, Virginia City Board of Aldermen,* October 15, 24, 1877, January 8, 1878; Interviews with William "Bud" Spargo, 1979–1980.

Chapter XI
Fighting Fire With Fire

1. *Gold Hill News*, June 24, 1872.
2. *Ibid.*, June 24, 25, 1872.
3. *Ibid.*
4. *Territorial Enterprise*, August 8, 1872; *Gold Hill News*, August 9, 1872.
5. *Gold Hill News*, August 19, 1872; *Territorial Enterprise*, December 2, 1872.
6. Clara Crisler, *History of the Warren Engine Company No. 1*, Carson City, Nevada; 1940. Private papers in the collection of Warren Engine Company No. 1, Carson City, Nevada.
7. *Gold Hill News*, February 6, 1873.
8. *Ibid.*, February 17, 1873.
9. *Ibid.*, February 6, 1873.
10. *Ibid.*, January 8, 1873.
11. *Ibid.*, January 14, 1873.
12. *Ibid.*, May 2, 1874.
13. *Ibid.*, May 25, 1874.
14. *Ibid.*, January 12, 1874.
15. *Minutes, Board of Fire Delegates, Gold Hill Fire Department*, June 23, 30, 1882, March 20, 1885; *Territorial Enterprise*, November 3, 1885; *Minutes, Virginia Board of Aldermen*, November 14–December 5, 1876; February 6, 13, 1877; Doten, p. 1296.
16. *Nevada State Journal*, October 15, December 18, 1903; *Minutes, Virginia Board of Aldermen*, May 14, 1878.

Chapter XII
A Deed Most Foul

1. *Gold Hill News*, November 18, 19, 1863.
2. *Ibid.*, November 13, 1868.
3. *Ibid.*, January 5, 1869.
4. *Ibid.*, May 5, 1869.
5. *Ibid.*, August 24, 1869.
6. *Ibid.*, August 28, 1869.
7. *Ibid.*, July 6, 1870.
8. *Ibid.*, November 25, 1870.
9. *Territorial Enterprise*, February 1, 1871; *Gold Hill News*, February 13, 1871.
10. *Territorial Enterprise*, February 14, 1871; *Gold Hill News*, February 13, 1871.
11. *Ibid.*
12. *Gold Hill News*, March 13, 1871; *Territorial Enterprise*, March 14, 1871.
13. *Ibid.*; *Grand Register of the Virginia Fire Department*.
14. *Territorial Enterprise*, March 15, 1871; *Gold Hill News*, March 14, 1871.
15. *Ibid.*
16. *Territorial Enterprise*, March 26, 1871; *Gold Hill News*, March 25, 1871.
17. *Gold Hill News*, April 5, 1871.
18. *Territorial Enterprise*, March 19, 1872; *Gold Hill News*, March 15, 16, 1871.

Chapter XIII
A Love-Hate Affair

1. *Ledgers*, Virginia & Truckee Railroad, Virginia & Truckee Railroad Museum, Nevada State Museum, Carson City, Nevada; *Gold Hill News*, May 5, 10, 1869; Doten, pp. 1043–1044, 1045.
2. *Gold Hill News*, May 10, 1869.
3. *Ibid.*
4. *Ibid.*, September 7, 1870.
5. Doten, p. 1304.
6. *Gold Hill News*, January 9, 1874.
7. *Minutes, Liberty Engine Company No. 1*, April 8, 1890.
8. A. M. Ardery, Diary, June 29, 1902. Virginia & Truckee Collection, Special Collections Department, University of Nevada-Reno Library, Reno Nevada.
9. *Ibid.*
10. *Ibid.*, November 17, 1903.
11. *Ibid.*
12. *Minutes, Liberty Engine Company No. 1*, February 9, March 8, 1904; W. H. Kirk, letters to H. M. Yerington, February 14, 17, 22, 23, 1904, Virginia & Truckee Collection, Special Collections Department, University of Nevada-Reno Library, Reno, Nevada.
13. *Ibid.*
14. *Ibid.*

Chapter XIV
Fire in the Mines

1. *Gold Hill News*, July 21, 1869.
2. *Ibid.*, October 30, 31, 1874.
3. *Ibid.*, October 11, 1875.
4. Lord, p. 270.
5. *Ibid.*, pp. 271–272.
6. *Ibid.*, p. 273.
7. *Ibid.*, p. 272.
8. *Gold Hill News*, April 8, 1869; Lord, p. 273.
9. Lord, p. 273.
10. *Ibid.*
11. Grant H. Smith, *History of the Comstock Lode* (Reno: University of Nevada Mackay School of Mines and the Nevada Bureau of Mines and Geology, 1943), pp. 122–123.

Chapter XV
Conflagrations

1. *Gold Hill News*, October 31, 1864.
2. *Ibid.*, March 13, 1869.
3. *Ibid.*, October 22, 1870.

4. *Territorial Enterprise*, February 1, 1871; *Gold Hill News*, February 1, 1871.
5. *Territorial Enterprise*, August 22, 1871; *Gold Hill News*, August 21, 1871.
6. *Gold Hill News*, March 6, 1873.
7. *Ibid.*, May 20, 1875.
8. *Ibid.*, June 30, 1875.
9. *Ibid.*, September 3, 1875.
10. *Ibid.*, October 6, 1875.

Chapter XVI
Virginia City's Great Fire of 1875

1. Wells Drury, *An Editor on the Comstock Lode* (Palo Alto, California: Pacific Books, 1936), p. 120.
2. Grant Smith, p. 176.
3. *Ibid.*, pp. 185–189, 191.
4. *Gold Hill News*, October 6, 31, 1875.
5. Doten, p. 1260; *Gold Hill News*, October 14, 1875.
6. *Gold Hill News*, October 26, 1875.
7. Dan De Quille, *The Big Bonanza* (New York: Alfred Knopf, 1947).
8. *Gold Hill News*, October 26, 27, 1875, March 9, 1876.
9. *Ibid.*, October 27, 1875.
10. De Quille, pp. 428–436.
11. *Territorial Enterprise*, April 9, 1954.
12. Ferdinand Beck, "Recollections of Early Virginia City," c. 1932, Greenhalgh Family Papers.
13. *Gold Hill News*, October 26, 27, 1875.
14. *Ibid.*
15. *Territorial Enterprise*, October 27, 1875; *Gold Hill News*, October 26, 27, 1875.
16. *Territorial Enterprise*, October 27, 1875; Drury, p. 120.
17. De Quille, p. 429.
18. Grant Smith, p. 192; Patrick Manogue, Diary of pledges and contributions to building fund of St. Mary's in the Mountains, 1875, Private Collection, Tucson, Arizona.
19. *Gold Hill News*, October 26, 27, 1875; *Territorial Enterprise*, October 27, 1875.
20. *Ibid.*
21. *Ibid.*
22. *Gold Hill News*, November 10, 1875.
23. *Ibid.*, October 26, 27, 1875; *Territorial Enterprise*, October 27, 1875; Drury, p. 120.
24. *Gold Hill News*, October 26, 1875, October 27, 1875; *Territorial Enterprise*, October 27, 1875.
25. *Gold Hill News*, October 29, 1875.
26. *Ibid.*, October 27, 1875.
27. *Ibid.*, October 26, 29, 1875; Lord, p. 328.
28. Lord, p. 328.
29. Beck, Recollections; *Territorial Enterprise*, November 25, 1955.
30. Beck, Recollections.
31. *Territorial Enterprise*, October 28, 1875; *Gold Hill News*, October 29, 1875; Grant Smith, p. 194.

Chapter XVII

Bitterness and a Paid Department

1. *Virginia Evening Chronicle*, November 1, 1875.
2. *Ibid.*, October 29, 1875.
3. *Gold Hill News*, November 9, 1875.
4. *Ibid.*
5. *Ibid.*
6. *Ibid.*, March 9, 1876.
7. *Territorial Enterprise*, March 5, 1876.
8. *Ibid.*, March 8, 1876.
9. *Gold Hill News*, July 5, 1876.
10. *Minutes, Virginia Board of Aldermen*, October 24, 31, November 14, 1876; *Territorial Enterprise*, October 25, 1876; *Minutes, Board of Fire Delegates, Gold Hill Fire Department*, December 20, 1876.
11. *Minutes, Virginia Board of Aldermen*, January 2, 1877.
12. *Regulations Governing the Virginia Paid Fire Department* (1877), Author's collection; *Statutes of Nevada* (Carson City, Nevada: State Printing Office, 1877), pp. 150–152.
13. *Gold Hill News*, March 10, 1877.
14. *Territorial Enterprise*, March 14, 1877.
15. *Ibid.*, March 7, 15, 18, 22, 1877.
16. *Ibid.*, March 25, 1877; *Virginia Evening Chronicle*, March 26, 1877; *Gold Hill News*, October 17, 1877.
17. *Territorial Enterprise*, March 21, 1877; *Minutes, Virginia Board of Aldermen*, August 14, 1877.
18. *Minutes, Virginia Board of Aldermen*, October 23, 1877.
19. *Ibid.*, May 14, 1878.
20. *Minutes, Virginia Exempt Firemen's Association*, November 28, 1876.
21. *Territorial Enterprise*, December 13, 1877.
22. *Ibid.*, July 2, 1878; *Minutes, Virginia Exempt Firemen's Association*, February 18, 1877; *Constitution and By-Laws, Virginia Exempt Firemen's Association* (Virginia City, Nevada: 1880), Nevada Historical Society, Reno, Nevada.
23. *Territorial Enterprise*, July 4, 1872, February 21, 1878; *Minutes, Virginia Exempt Firemen's Association*, February 18, 1877.
24. *Territorial Enterprise*, July 1, 1878.
25. *Ibid.*, March 24, 1883.
26. *Report of the Chief Engineer, Virginia Paid Fire Department*, June 9, 1883, Storey County Clerk's Office, Storey County Courthouse, Virginia City, Nevada.
27. *Ibid.*, 1883–1896.
28. *Minutes, Gold Hill Fire Department*, 1880–1990.
29. *Ibid.*, November 15, 1885.
30. *Territorial Enterprise*, October 3, 1885.
31. *Ibid.*, July 27, 1888.
32. *Virginia Evening Chronicle*, July 5, 8, 1890.
33. *Ibid.*, July 18, August 23, 1890.
34. *Ibid.*, August 4, 1890, November 1, 1892.
35. *Territorial Enterprise*, August 17, 1892.

36. *Ibid.*, August 23, 1890.

37. *Virginia Evening Chronicle*, July 6, 1891.

38. *Ibid.*, April 2, 1896.

39. *Ibid.*, April 2, 1886; Interview with Alice Byrne, 1980.

40. *Reports of the Chief Engineer, Virginia Paid Fire Department*, 1901–1906.

41. Datin, pp. 41–46.

42. *Nevada Appeal*, January 8, 1978.

43. *Virginia Evening Chronicle*, January 26, 1900; *Minutes, Gold Hill Fire Department*, 1867–1896.

44. Interview with Vic Maxwell, November 11, 1977; Interviews with Delbert Benner, 1980.

45. *Nevada State Journal*, November 13, 14, 1942; Interview with Walt Mulcahy, October 29, 1982; Interview with John Guiffra, October 31, 1982; *Fire Book, Storey County Volunteer Fire Department*, November 13, 1942, Comstock Firemen's Museum, Liberty Engine Company No. 1, Virginia City, Nevada.

46. *Nevada State Journal*, August 30, 1944; Mulcahy interview.

47. *Minutes, Storey County Volunteer Fire Department*, 1952–1954, Comstock Firemen's Museum, Liberty Engine Company No. 1, Virginia City, Nevada.

Doten Sidebar

1. *Grand Register of the Virginia Fire Department; Roll Book, Yellow Jacket Engine Company No. 2*, Comstock Firemen's Museum, Liberty Engine Company No. 1, Virginia City, Nevada; *Minutes, Yellow Jacket Engine Company No. 2*, Nevada Historical Society, Reno, Nevada.

2. Doten, p. 906.

3. *Ibid.*, p. 910.

4. *Ibid.*, pp. 940–941.

5. *Ibid.*, pp. 1292, 1294, 1296, 1299.

6. *Ibid.*, pp. 1321–1322.

7. *Minutes, Virginia Exempt Firemen's Association*, Nevada Historical Society, Reno, Nevada.

Marks Sidebar

1. *Minutes, Gold Hill Fire Department*, 1871; *Gold Hill News*, April 24, 1873.

2. *Gold Hill News*, October 15, 1874.

3. *Minutes, Gold Hill Fire Department*, May 18, 25, 1873.

4. *Ibid.*, June 1, 1883.

5. *Ibid.*, June 15, 1883.

6. *Ibid.*

7. *Minutes, Liberty Engine Company No. 1*, 1887–1895. Comstock Firemen's Museum, Liberty Engine Company No. 1. Virginia City, Nevada; *Minutes, Gold Hill Fire Department*, 1867–1896.

Peasley Sidebar

1. Angel, p. 599.

2. *Ibid.*, p. 266.

3. *Gold Hill News*, June 23, 1865; *Journal, Senate of the State of Nevada, 1864–1865* (Carson City, Nevada: Legislative Counsel Bureau Library), p. 7.

4. Katie Hillyer and Katie Best, *Julia Bulette & Other Red Light Ladies* (no location: Western Publications, 1975), pp. 26–27.

5. *Grand Register of the Virginia Fire Department.*

6. *Ibid.*

7. *Gold Hill News*, January 9, 1865.

8. Anthony Amaral, *Lace Curtains and Bootjacks* (Carson City, Nevada: n.p., 1975), p. 10; Joseph Goodman, *Heroes, Badmen and Honest Miners* (Reno, Nevada: Great Basin Press, 1977), pp. 43–48.

9. Goodman, pp. 43–48.

10. Doten, pp. 876–877.

11. *Ibid.*

Tweedie Sidebar

1. *Grand Register of the Virginia Fire Department* and *Membership Roll, Monumental Engine Company No. 6*, Comstock Firemen's Museum, Liberty Engine Company No. 1, Virginia City, Nevada.

2. Interviews with Delbert Benner, Alice Byrne, Margaret Marks, William Spargo, 1979–1981.

3. *Ibid.*

4. *Ibid.*

5. *Ibid.*

6. Interview with Ron Rice, nephew of Dave Tweedie, Benicia, California, 1981.

Index

Italic page numbers refer to photographs.

Adkison, D. O., 7
Aine, Henry, 159
Alchorn, Isabella, 42
Alchorn, Thomas, 185, 190–191
Aldermen, Board of, 17, 18, 64, 94, 113
American Exchange, 163–164
American Saloon, 217
Andrews, Thomas, 5
Antunovich, Andy, 217
Ardery, A. M., 148, 150
Ashland House, 29
Assembly Saloon, 177
Athletic Hall, 39, 40, 144
Atkinson, T. A. (sheriff), 144
Aylsworth, G. W., 78

Babcock Engine Company No. 1 (Bodie, Calif.), 87
Baglin, Henry, 16, 209
Baker, J. W., 168
Ball, N. A. H., 78
Ballou, Benjamin, 25, 80
Banaham, Mary, 118
Bank Exchange Saloon, 87
Banner Brothers, 180
Barnhart, Martin V., 88–91
Barry, ____, 26
Barstow, William H., 5
Bartlett, Billy, 204
Bassett, May, 156
Bateman, I. C., 5
Battu's Paint Shop, 164
Beck, Ferdinand, 174, 182–183
Beebe, Libbie, 196
Beegan, Robert, 31
Belcher Mine, 30–31, 155–156
Belknap, C. H., 193–194
Bell, Joseph P., 16, 38, 136
Belvedere Hotel, 29
Benner, Delbert, 214, 220
Bickle, George, 157–158
Bickle, Richard, 157–158
Birdsall, George, 16, 17, 111
Bittner's Hardware Store, 36
Black & Brothers, 141, 177
Bob, Eddie, 217
Bodman, J. C., 140
Bowers Grade, 131
Bowers Mansion, 146, 208
Bowers Mill, 66
Bray, Flora, 52
Breed, ____, 26
Brennan, W., 136
Brewer, Maggie, 52
Brokaw, I. E., 5
Brown, James K. B. "Kettlebelly," 16, 189–191, 198
Brown, Richard, 197
Bryan, Charles, 111
Bryarly, Dr. ____, 26
Bulette, Julia, 54–56, 66, 88
Button, Lysander, 133
Byrne, Alice, 210
Byrne, John, 210

California Consolidated Mining Co., 170, 177, 236
Capitol Saloon, 80
Carson Brewery, 177
Carson City Fire Department, 127
Carson Daily Appeal, 72
Cartter, James N., 38
Central Mill, 27
Central Pacific Railroad, 133, 146–147
Chinatown, 98, 112–113, 175, 179
Chollar-Potosi Mining Co., 154–155
City Bakery, 168
City Cemetery, 55
City Hall, 172
City Hose Cart No. 1, 127–128, *196*, 210, 212, 213, 214, 217, 219, *220*, 221
City Hose Cart No. 2, 127, 212, 213
Clark, Carrie, 42
Clarke, Rev. ____, 35
Clemens, Jule, 25
Clemens, Samuel, 1–3, 109

Cobb, Ty, 217
Coiture's Grocery & Variety Store, 34
Cole's Drug Store, 199
Collins House, 142
Columbia House, 177
Combination Shaft, 215
Commercial Soap Works, 205
Comstock, A. V., 7, 190–191
Comstock Firemen's Museum, 67, 68, 123, 128, 218
Confidence Engine Company No. 6, 102, 222
Conlan, Frank, *196*, 212
Consolidated Virginia Mining Co., 170, 177, 214, 236
Cook, Owen, 30
Cooper's Hall, 198
Corcoran, Daniel, 172
Cornellier, Desire, 156
Corner Bar, 90
Corporation Fire House, 68, 113–114, 127, 193, *194*, 198, 204, 207, 210, 217, 219–220
Corrales family, 217
Coryell, Special Officer ______, 207
Coyle, Tom, 215
"Crazy Kate." *See* Shay, "Crazy Kate"
Crisler, Clara, 48–49
Crocker, Charles, 146–147
Crocker, H. S. & Co., 8
Crocker, Mollie, 214
Cronin, Tim, 27
Crosby, D., 102
Crossman, John, 191
Crown Point Mine, 30–31, 157–161
Cullen, John P., 84–87
Cummings, W. I. (sheriff), 25, 35, 165–166
Curry Engine Company No. 2 (Carson City, Nev.), 77, 81, 142, 208

Daley, Alex, 173–174
Daley, George, 173–174
Darwin's Assay Office, 141
Davenport, William H., 69
Davis, Fred, 212
Dayton Fire Department, 99–100, 215
Deidesheimer, Phillip, 7
Deininger, John, 197
Delegates, Board of Fire (Gold Hill), 69, 70–71, 73–75, 124–126
Delegates, Board of Fire (Virginia City), 16–17, 70
Delta Saloon, 177, 197
Denning, Lizzie, 42
De Quille, Dan. *See* Wright, William
Derby & Gearhart's Stables, 177
DeWitt Clinton Commandery No. 1, Knights Templar, 168
Dias & Blauber Building, 32
Divide Hose Company No. 2, 52, 94, 123, 124, *125*, 126, 128, 200, *201*, 202, 204–205, *206*, 207, 209, 210, 212, 213–214, 218, 222
Doten, Alfred, 7, 13, 18, 20, *21*, 23, 25, 27, 39, 55, 80, 90, 93, 98, 134, 145, 146, 184, 204
Downey, George, 48, 112–113, 143, 165
Drury, Wells, 7, 20, 170, 176
Dumar's Saloon, 79
Dunlop, Johnny, 203
Dunn, John, 65
Dustan, Mrs. A. T. *See* Brewer, Maggie

Eagle Engine Company No. 3, 20, 27, 29, 35, 42, 45, 46–47, 52, 54, 57, 60, 63, 76–77, 80, 81, 84, 88, 90, 96, *97*, 98, 115–116, 163, 168–169, 172, *173*, 175, 179, 185, 186, 189, 190, 195–196, 222, 224–235, 236–237
Easterbrook, H. H., 107
E Clampus Vitus, 55
Eclipse Mill, 36, 38
Egan, John, 204
Emerson, Katie, 196–197
Emmett Guard, 168, 174
Engine Companies. *See* Confidence Engine Company No. 6; Curry Engine Company No. 2; Eagle Engine Company No. 3; Eureka Engine Company No. 1; Howard Engine Company No. 3; Knickerbocker Engine Company No. 1; Knickerbocker Engine Company No. 5; Liberty Engine Company No. 1; Monumental Engine Company No. 6; S. T. Swift Engine company No. 3; Vigilant Engine Company No. 9; Virginia Engine Company No. 1; Warren Engine Company No. 1; Washoe Engine Company No. 2; Washoe Engine Company No. 4; Yellow Jacket Engine Company No. 2; Young America Engine Company No. 2
Ennis, George, 5
Ennis, Pat, 174
Escurial Lodge No. 7, 168
Eureka Engine Co. No. 1 (Marysville, Calif.), 99
Eureka Saloon, 162

Fair, James G., 42, 132, 197
Fair, Jimmy, 42
Farrington, F. C., 20

Ferrend, Major ____, 32, 168–169
Ferris, Judge Leonard W., 109
Fireman's Trust Bank, 120
First Ward School, 177, 179
Fish, Charles, 5
Flick, Philip, 118
Flower, Joseph, 120
Flurshutz's Confectionery Store, 144
Fogg, G. A., 5
Folsom & Hiller, 108, 119
Fourth Ward School, 200, 215
Fox, Frank, 209
Fox, Thomas H., *14*, 16
Fredericks, Joseph, 174
Fredericks's General Store, 174
Fredrick, M. M., 177
Fredricksburg Brewery, 174
French Rotisserie, 205
Friedlander, Gustav, 200
Fulton Market, 177

Gallagher, Thomas, 135
Garavanta, Julio, 215
Garden Valley House, 26
Gardner, Abigail, 182–183
Gardnerville Fire Department, 94
Garvey, Bruce, 27
George, J., 116
Gibson, Amalia, 52
Gibson, W. D. C., 16, 136
Gillette, Col. ____, 196
Gillig, Mott & Co., 70, 118, 177
Gillis, Steve, 20
Giuffra, Enrico, 217
Gladding, Bessie, 217
Gladding, J. F., 63, 124, 200
Golden Eagle Hotel, 162–163
Gold Hill Brass Band, 73
Gold Hill Fire Department, 5, 7, 13, 16, 20, 30, 33, 36, 49, 52, 63, 70–71, 73–75, 78, 81, 119, 124–126, 132, 136, 150, 175, 200, 209, 210
Gold Hill Market, 36
Gold Hill News, 7–8, 13, 18–22, 24, 26, 29, 30, 31, 35–36, 38, 40, 42, 45, 52, 59, 64, 66, 76, 77, 79, 81, 87, 93, 95, 96, 98, 99, 102, 104, 106, 112, 116, 119, 120–121, 131, 133, 134, 140, 141–142, 144–145, 146–147, 155, 171, 175, 179, 180, 184–185, 188, 191
Goodman, Joseph, 7, 20
Good Will Hose Company No. 2, 42, 115, 121, *122*, 147, 189, 222
Good Will Hose Company No. 25 (Philadelphia, Pa.), 121, *122*
Goodwin, C. C., 176
Gould & Curry Mining Co., 26, 44, 163, 183, 198–199, 202
Graham, A. J., 79
Granite Mill, 38, 78
Graves, Robert N., 193
Greer, Addie, 52
Greiner, William, 49, 214
Grosetta, Martin, 5
Gumbert & Webber, 197

Hale, W. E., 120
Hale & Norcross Mine, 155
Hamburger, Jules, 27
Hanbridge, George W., 200
Hand-In-Hand Hose Company No. 2, 222
Happs & Finnery Paint Shop, 26
Hardy, R. C., 5
Harrah, William F., 100
Hart, G. A., 120
Hatch, George, 203
Haynie's Lumber Yard, 177
Heilshorn, Richard, 167
Hemenway, J. W., 195
Higbee, B. L., 165
Higgins, Mrs. Thomas, 215
Hiller, Dr. Frederick, 142
Hillyer, M. C., 5
Hirschman, A., 5
Hobson, Frank, 215
Holland, Mamie, 215
Holman, Heber "Hebe," 35, 202
Holman, Hi, 31
Holsinger, Fred, 102
Hopkins, G. W., 7
Hose Companies. *See* City Hose Cart No. 1; City Hose Cart No. 2; Divide Hose Company No. 2; Good Will Hose Company No. 2; Good Will Hose Company No. 25; Hand-In-Hand Hose Company No. 2; Invincible Hose Company No. 4; Junior Hose Company No. 1; Liberty Hose Company No. 1; Lincoln Hose Company No. 3; Neptune Hose Company No. 5; Our Own Hose Company No. 5; Rescue Independent Hose Company; Rooster Hose Company No. 1; Silver City Hose Company No. 1; Yellow Jacket Hose Company No. 2
Howard Engine Company No. 3 (San Francisco), 77, 102–103, 147
Hubbs, David, 116
Huddy, Tom, 214
Humphrey, David, 31

Imperial-Empire Mine, 30, 168
International Hotel, 5, 95, 98, 126, 168,

171, 174, 207, 210–213, *211*, 236
Invincible Hose Company No. 4, 115, 116, 118, 143–145, 222

Johns, William, 156
Johnson, Elwood, 217
Jones, Abe, 16, *28*, 36
Jones's Apothecary, 36
Jones, John, 7, 135
Junction House, 162
Junior Hose Company No. 1, 115, 123, 222

Kaneen, J. S., 7, 190–191
Kaplan, L., 22
Kelley, Patrick, 35, 156
Kelly & Lowenstein's Store, 32
Kennedy, Michael, 16
Kennedy & Mallon's Grocery, 177
Kenney, John, 212
Kennis, John, 67
Kentuck Mine, 30–31, 157–161
Kerrin, Hugh, 13, 16
Ketton, James, 174
Kimball Manufacturing Co., 127, 188
King, George, 212
Kirk, William, 150–153
Kline, Lisa, 214
Knickerbocker Engine Company No. 1 (Dayton), 99
Knickerbocker Engine Company No. 1 (Sutro), 100
Knickerbocker Engine Company No. 5, 16, 22, 27, 34, 36, 42, 46–47, 57, 64, 81, 92, 99, *100*, *101*, 102, 105, 113, 115, 119, 133, *134*, 135, 136, 163, 174, 177, 179, 181, 186, 189–190, 194, 195, 213, 222, 236–237
Knox, Judge ____, 172

Laird, Robert, 156–157
Lambert, John, 35
Lammon, George I., 5
Larkin, Peter (chief), 13, 38, 69–70, 88, 93
Larkin, Peter (murderer), 172
Larrowe, C. W., 162
Laswell, Thomas, 144–145
LeClare, Frank, 156
Leconey, John, 29
Lee, John T., *9*
Lee, W. H. H., 59, 77, 120, 159
Liberty Engine Company No. 1, 8, 29, 30, 31, 35, 40, 45, 46–47, 48–49, 52, *58*, 59, *60*, 66, 70–72, 73–75, 76–79, 80–83, 92, 102–107, *105*, 117, *122*, 123–124, 131, 146–153, 159, 167, 200, 212, 213–214, 218, 222
Liberty Hose Company No. 1, 27, 31, 49, 115, 117, 119, 123–124, 128, *160*, 209, 222
Lincoln Hose Company No. 3 (Gold Hill), 8, 63, 67, 70, 74, 81, 104, 123, 222
Lincoln Hose Company No. 3 (Virginia City), 116, 189, 222
List, Gov. Robert, 128
Locke, Judge P. B., 86
Lone Star Stables, 204–205
Lonkey's Wood Yard, 207
Lord, ____, (captain), 190–191
Lord, Eliot, 157, 159, 181
Lutgens, Otto, 70
Lynch, Patrick, 84
Lynch's Saloon, 84
Lynds, J. B., 204–205
Lynds's Grocery, 215
Lynds's Stables. *See* Lone Star Stables
Lyon, W. Parker, 94, 100

McCausland, John, 177, 199
McCone, Dolly, 52
McCone, Mrs. John, 177
McCone, Susie, 52
McDermott, Charles, 209
McDonnell, John F., 136
McGinnis, Terence, 5
McGowan, Mike, 27
McKay, James, 190–191
Mackay, John, 67, 176–177, 183
McKay's Stables, 165
McMartin, Alex "Sandy," 16, 30, 35, 74, 79
McNair, Frank, 16, 184, 187–188, 189–191
McQuigan, Charley, 212–213
McWilliams, Charles, 144–145
Magnolia Saloon, 177
Malone, James, *11*, 16, 38
Malone, Mike, 174
Mann, Mamie, 52
Manogue, Rev. Patrick, 158, 176–177, 188
Markey, Eugene, 118
Marks, John, 16, 31, 52, 73–75, 79, 200
Marks, Meriam, 52, 75
Marye Building, 177
Marysville (Calif.) Fire Department, 99
Masonic Lodge. *See* DeWitt Clinton Commandery No. 1, Knights Templar; Escurial Lodge No. 7; Virginia Chapter No. 3 of R.A.M.; Virginia Lodge No. 3
Maxwell, Vic, 214

May, George, 5
Meagher, John, 152
Menken, Adah Isaacs "La Menken," 54
Mercer, Richard, 159
Merchants' Exchange, 165
Mesick, R. S., 80
Metcalf, M. W., 63, 83
Methodist Church (Gold Hill), 63
Methodist Church (Virginia City), 177
Mexican Mining Co., 5, 142
Milatovich's, 142
Millain, John, 55–56, 66
Miller, Pierce A., 127, 213
Milligan, William, 5
Milton House, 57
Miners' League, 162
Miners' Union (Gold Hill) 40, 136, 148, 208
Miners' Union (Virginia City), 113, 177
Montgomery Guard, 168
Monumental Engine Company No. 6 (San Francisco), 66, 84, 95
Monumental Engine Company No. 6 (Virginia City), 29, 34, 36, 44, 46–47, 57, 66–67, 102, 115, 135, 136–139, *138*, 147, 156, 164, 167, 172, 174, 179, 186–187, 189–190, 199–200, 219, 222, 236
Mooney's Livery Stable, 177
Moor, John, 2
Mount Davidson House, 162
Mueller, Charley, 197
Mulcahy, Pat, 31–32
Murphy, John, 157
Music Hall, 93

Nagle, William, 215
Neptune Hose Company No. 5, *12*, 42, 44, 115, 118, 119, 167, 189, 222, 236
Nevada Artillery, 168
Nevada Brewery Saloon, 29
Nevada Hook and Ladder Company No. 1, 5, 16, 52, 57, 84–86, 87, 108–114, *110*, *112*, 143, 159, 165, 179, 186, 188, 189, 194, 195, 213, 222, 237
Nevada Legislature, 16, 31, 88, 189
Nevada National Guard, 168, 179. *See also* Emmett Guard; Montgomery Guard; Nevada Artillery
Nevada State Grazing Division, 215
Nevada State Journal, 48, 139, 217
Nevada State Museum, 127–128, 217
Nevada State Prison, 86–87, 145, 167
Nevin, Michael E., 210
New York Fire Department, 83, 109
Niagara Saloon, 5
Nolan, Billy, 27
North, Judge J. W., 86–87
Nuttall, Charles E., 79
Nutter, C. T., 215
Nye, Harry, 207–208
Nye, Peter, 59

Occidental Lodging House, 169
Odd Fellows' Lodge, 40, 168–169, 205
"Old Butt," 31–32
"Old Dave," 197
"Old Gray," 213
O'Neil, W. T., 141–142
Ophir Mine, 5, 144–145, 170, 177, 181, 236
Ormsby House, 88–90
Oro Hotel, 39
Orrum, Eilley, 66
Our Own Hose Company No. 5, 115, 222
Overman Mine, 214

Pacific Brewery, 172
Pacific Lodging House, 202
Paddock, Richard, 25
Panian, Martin, 29–30
Pappas, Mandy, 217
Parke, Frank, 196
Parker, James, *14*
Paster, Sam, 5
Paster's Saloon, 5
Peabody, ____, 35
Pearl, Emie, 215
Pearson, Bill, 215
Peasley, Andrew, 88, 197
Peasley, Thomas, 13, 25, 54, 80, 87–91, *89*, 109–111, 195, 197
Pennison, William, *15*, 16, 190–191, 195, 198, 200, 204
Perkins, Arthur, 143–145
Perry, John Van Buren, 109–111
Phelan, James, 7, *14*, 29, 44, 52
Philadelphia Brewery, 36, 177
Philadelphia Fire Department, 121
Piper, John, 7, 198
Piper's Opera House and Saloon, 39, 52, 57, 80, 142–144, 165, 177, 197–198, 212, 236
Pittman, John, 215
Plunkett, Fred, 23
Pohl, Coonie, 49
Pohl, Florence, 49
Pohl, Minnie, 49, *50*, 51
Pollard, Richard, 71, 156
Porteous, Sam, 203
Potter, ____, 143
Pottle, A. E., 190–191

Presbyterian Church, 63, 129
Price, John, 106
Provost Guard Station, 27, 144
Pruitt, Mr. & Mrs. ____, 182
Putnam, A. A., 35, 81, 159

Ralston, William C., 170
Red Men Lodge, 169
Reed, John, 5
Reese, Ed, 35
Reim's Saloon, 177
Reno Fire Department, 99, 186
Reno Police Department, 215
Rescue Independent Hose Company, 189, 222
Rheim, Matt, 214
Rhode Island Mill, 63, 135
Richardson, Edward, 84–87
Riorden, John, 198–200
Rising, Judge Richard, 145, 177
Ritchie's Barbershop, 203
Romano, Felix, 214
Rosenbaum's Furniture Store, 202–204, *202*
Roos Brothers Emporium, 177
Rooster Hose Company No. 1, 26, 42, 115, 189, 222
Rower, Billy, 35
Rule, James G., 202
Rush, John, 198

Safford, Charles, 27
St. Louis Brewery, 26, 141
St. Mary's in the Mountains, 132, 158, 174–175, 176–177, 183, 236
St. Paul's Episcopal Church, 141, 183
San Francisco Fire Department, 83, 84, 95–96, 102, 147
San Francisco Restaurant, 36
Savage Mine, 196, 199
Sazerac Saloon, 48
Scannel, David, 120
Scholl's Gunsmith Shop, 164
Schwab, George, 168
Scott, A. C., 163–164
Scott, Gussie, 42
Scullion, John, 174
Sease, Bernard, 127–128
Sexsmith Building, 212
Sexton, Nick, 67
Sharon, Amelia, 214
Sharon, Frank, 215
Sharon, William, 214
Shaw, George, 5
Shay, "Crazy Kate," 171, 236
Sheehan, Jerry, 31
Sheppard, William, 80
Sheridan, Gen. Phillip, 171
Sherman's Lodging House, 165
Silver Bar Hook and Ladder Company No. 1, 119, 222
Silver City Hose Company No. 1, 81
Sims, Etta, 52
Singleton Lodging House. *See* Pacific Lodging House
Smith, Emory, 207–208
Smith, George, 64
Smith, J. P., 141
Smith, John, 5
Spargo, William "Bud," 127–128, 219
Spirit of the Times, 120
Sprague, Mrs. P. M., 182
Stack, Tim, 71–72, 106
Stapleton, James, 5
Star House, 93
Stearns, David, 29
Stoner, Philip, 144
Storey County Commission, 74, 127, 210, 236
Storey County Courthouse, 57, 123, 171, 180
Stout, William, 143
S. T. Swift Engine Company No. 3 (Carson City, Nev.), 133
Sugar Foot Jack, 88
Sullivan, Maggie, 118
Summit Lake House, 140
Sutro, Adolph, 196
Sutro Fire Department, 100, *101*, 102
Swift, E. C., 150
Swift, Shubal T., 133
Symons, Humphrey, 180

Taylor, Willie, 167
Territorial Enterprise, 13, 20, 66, 73, 95, 99, 102, 119, 121, 131, 173, 175, 176, 177, 183, 187–188, 191, 193, 200, 203, 204, 236
Theatre Hall, 45
Third Ward School, 179
Thomas, C. C., 204
Thompson, Cad, 93
Thompson & Metcalf, 63
Three Bridges, 167
Tobener, Charles, 71
Tritle, F. A., 121, 147
Twain, Mark. *See* Clemens, Samuel
Tweedie, Dave, *196*, 212, 219, *220*, 221
Tweedie, John, 215, 221
Twoomy, ____, 152

Uniforms, 7–9
Union Consolidated Tunnel, 172
United States Army, 88, 109, 111, 215
Upper Level, 141
Utah Mine, 156–157

Valentine, ____, 151
Vesey House, 29, 35, 140
Vigilant Engine Company No. 9 (San Francisco), 96
Vigilantes ("601"), 66, 143–145
Virginia Chapter No. 3 of R.A.M., 168
Virginia Consolidated Mining Company, 167
Virginia Engine Company No. 1, 3, 5, 16, 17, 18, 26, 27, 38, 42, 54–56, 67–68, 76, 84–86, 87–88, 93–95, 109, 115, 119, 163, 179, 186, 189, 191, 194, 195, 204–205, 222, 237
Virginia Evening Bulletin, 96
Virginia Evening Chronicle, 177, 184–185, 191, 207, 209, 236
Virginia Exempt Firemen's Association, 16 22, 45, 94, 195–197, 218
Virginia Fire Department, 3, 5, 7–8, 13, 18, 22, 23, 24, 38, 48, 69–70, 84, 88, 96, 119, 121, 132, 136, 167, 169, 171, 179, 184–194, 200, 204, 237
Virginia Hotel, 177
Virginia House, 199
Virginia Paid Fire Department, 16, 94, 126, *127*, 136, 150, 189–195, *192*, *194*, 195, *196*, 200, 208–210, *208*, 219–221
Virginia Saloon, 5
Virginia & Truckee Railroad, 77, 81, 102, 107, 131, 133, 146–153, *149*, 175, 177, 180, 182, 214, 217, 236
Vucovich, Spiro, 197

Wade, Ben, 215
Wadge, John, 70
Waider, C. J., 71
Walker, J. M., 5
Warnock, Billy, 86, 163
Warren Engine Company No. 1 (Carson City), 46, 48–49, 77, 81–83, 88–90, 127–128, 146, 208, 214, 215, 216
Washington Hose Company No. 1 (San Francisco), 120
Washington Market, 5
Washington Saloon, 177
Washoe County Sheriff's Department, 215
Washoe Engine Company No. 2 (Reno), 139
Washoe Engine Company No. 4, 20, 22, 38, 39, 42, 46, *47*, 57, 72, 80, 84–86, 96–99, 115, 118, *132*, 133, 136, 143, 145, 147, 164–165, 172, 184, 186, 189, 195, 222
Washoe Saloon, 36
Washoe Stables, 164
Waters, James, 5
Webster, Paul, 216
Weigand Assay Office, 36
Weir, William, 16
Wells, Fargo & Co., 26, 163–164
Western Union Telegraph Co., 168, 175
Wetherill, Sam, 5, 140
White, Jake, 40
Wiley, Martin, 168
Williams, Lizzie, 52, 111, 163
Williams, M. R. "Riff," 13, 16, 24–25
Williams, O. D., 174
Willis, William, 143–145
Willock, James C., 13
Wilmot, Peter, 168
Wilson, Katy, 52
Wilson & Zoyara's Circus, 45, 119
Winterbauer, J. A., 70–71, 74
Wolf, M. P., 79
Wood's Cigar Stand, 13
Wrenn's Saloon, 213
Wright, S. D. "Wanie," 13, 16, 26
Wright, William (Dan De Quille), 171, 172, 176
Wyckenheim, Sam, 64, 144
Wylie, ____, 39, 164–165

Yellow Jacket Engine Company No. 2, 8, 48–49, *59*, 74, 77–79, *117*, 123, 124, *125*, 135–136, *137*, 147, 159, 175, 189, 200, 222, 237
Yellow Jacket Hose Company No. 2, *6*, 20, 36, 76, 120, 146, 222
Yellow Jacket Mine, 24, 30–31, 157–161
Yerington, Henry M., 133, 148, 150–153
York, Carl, 40
Young America Engine Company No. 2, 16, 26, 29, 34, 38, 40, 42, *43*, 44, 45, 46–47, 52, 54, 56, 57, *62*, 63, 64, 65, 66, 84–87, 94–96, 98, 115, 121, 129, *130*, 131–133, 136, 144, 154, 163–164, 166, 168, 169, 186, 189–190, 222
Young Brothers Saloon, 203
Young, Charley, 212, 215
Young, Jacob, 7, 16

Zouave Mining Co., 5

ABOUT THE AUTHOR

Steve Frady, fire chief of the Virginia City Volunteer Fire Department, has been actively involved in preserving the remaining apparatus of the old Virginia Fire Department and Gold Hill Fire Department. He is one of the founders of the Comstock Firemen's Museum, which is dedicated to preserving not only the equipment and memorabilia of the old fire companies, but also their traditions. Frady, an award-winning journalist, has spent more than seven years traveling throughout the West to research fire department records, diaries, and minutes and to unearth previously unpublished photographs for this book.